Annals of the ICRP

ICRP PUBLICATION 86

Prevention of Accidental Exposures to Patients Undergoing Radiation Therapy

Editor
J. VALENTIN

PUBLISHED FOR

The International Commission on Radiological Protection

UK Elsevier Science Ltd, The Boulevard, Langford Lane, Kidlington, Oxford OX5 1GB, UK

USA Elsevier Science Inc., 660 White Plains Road, Tarrytown, New York 10591-5153, USA

JAPAN Elsevier Science Japan, Tsunashima Building Annex, 3-20-12 Yushima, Bunkyo-ku, Tokyo 113, Japan

---

Third edition 2000
ISBN 008 044 0827
ISSN 0146-6453

Published quarterly (March, June, September, December)
September 2000 issue

Typeset by Variorum Publishing Ltd, Rugby
Printed and bound by CPI Group (UK) Ltd, Croydon, CR0 4YY

MIX
Paper | Supporting responsible forestry
FSC www.fsc.org FSC® C013604

Prevention of Accidental Exposures to Patients Undergoing Radiation Therapy

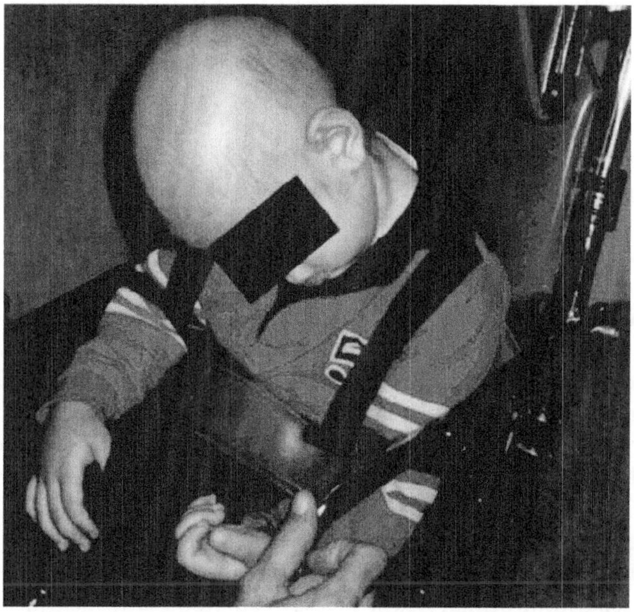

Fig. 1. Radiation overdosage to the brain, as a result of which this child lost his ability to speak and walk and has permanent epilation.

A publication from the
International Commission on Radiological Protection.

ICRP Publication 86

Annals of the ICRP

Editorial

IT HAS TO COME FROM WITHIN

It's a question that we all face now and then, in our private lives as well as professionally: how can we achieve safety?

And it's tempting, but rather futile, to come up with replies that essentially translate as 'by coercion'.

Of course, a few strict rules are a necessity; there must be enforcement in terms of licensing, inspection, and at times penalties for non-compliance. Exasperated young parents, trying to prevent accidents to their small children, may feel that there is no alternative to explicit bans coupled with threats of dire punishment (and, if need be, execution of such punishments).

Unfortunately, this approach is not very efficient, and as the children grow older, all sensible parents will want to instil a sense of personal responsibility in their children's minds, such that safety is the result of the children's own wishes and considerations, not the result of rules set by parents or of fear of punishment. As adults, we avoid illegal and hazardous behaviour, not because there might be a policeman round the corner catching us, but because it actually makes sense to behave safely.

Applying this parable to radiation safety, we can never achieve true safety by having 'sufficient numbers of inspectors' or 'sufficiently severe penalties' to keep unreliable operators on a short leash. Safety must come from within, from a genuine desire by operators to run a safe operation.

In such a scenario, a primary task of the regulator's is to help licensees by providing as much information as possible about the prevention of accidents. This is one of the important messages in the present publication: an incident reporting system with feedback to operators is vital in order to prevent the recurrence of accidents.

For such a system to work, mutual trust is required. Obviously, regulators need to be able to trust operators. In a sense, that's what licensing is about: operators must prove that they have the knowledge, the organisation, and the resources (including an internal audit mechanism) required to run a safe operation, and the license once issued is an indication of trust from the regulator. But it also has to work the other way around: operators must be able to trust regulators.

One aspect is that all operators need to be sure that the regulator will only award licenses to trustworthy operators - otherwise, serious operators would be unfairly disadvantaged. Another aspect, perhaps less obvious at first sight, is that operators must be confident that the regulator really is there primarily to achieve safety, not to find and punish culprits. This is particularly evident in the case of incident reporting. If the inevitable result of incident reporting is that the operator is punished, then few operators will be willing to come forward with their problems, and in the end safety will suffer for lack of information transfer.

Probably the best proof of this comes from aviation, where there is a long history of incident reporting. In aviation, operators and their staff are trained to volunteer information, not only about accidents, but also about unsafe situations that could have developed into an accident but did not go that far. They provide this incident information, convinced that the authorities will not want to press charges or punish them other than in cases of blatant and deliberate disregard for rules and procedures. The important point here is that the system of mutual trust has generated an unsurpassed level of safety that makes aviation the safest mode of transport by far.

Safety experts sometimes speak of four stages of safety awareness. Operators at the lowest stage simply try to evade authorities and rules. The next stage is when the operator complies with regulations, but tries to minimise the short-term efforts and costs. The third stage is when the operator is willing to spend any amount of effort and money to achieve safety, but requests the regulator to tell them exactly what to do. The highest level of safety awareness is when the operator is not only willing to invest whatever it takes, but also actively participates in furthering the safety objective, and is capable of conducting discussions with peers and regulators on an equal footing in order to further improve safety.

At least some of the accidents described in the present report would probably not have happened, or at least could have had less devastating consequences, if all operators had reached the highest stage of safety awareness. If this report can contribute to increasing safety awareness, then it will have achieved its objective.

JACK VALENTIN

CONTENTS

PREFACE

Over the years, the International Commission on Radiological Protection (ICRP), referred to below as 'the Commission', has issued many reports providing advice on radiological protection and safety in medicine. Its *Publication 73* is a general overview of this area. These reports summarise the general principles of radiation protection and provide advice on the application of those principles to the various uses of ionising radiation in medicine and biomedical research.

Most of these reports are of a general nature, and the Commission wishes to address some specific situations where difficulties have been observed. It is desirable that reports on such problem areas be written in a style which is accessible to those who may be directly concerned in their daily work, and that every effort is taken to ensure wide circulation of such reports.

A first step in this direction was taken at the Commission's meeting in Oxford, United Kingdom, in September 1997. At that time, on the recommendation of ICRP Committee 3, the Commission established several Task Groups to produce reports on topical issues in medical radiation protection.

Some such reports have now appeared in print, and the present one is the result of the work of yet another one of those Task Groups. The terms of reference of this Task Group on Accident Prevention and Safety in Radiation Therapy were to prepare a document with recommendations to prevent accidents. The Task Group was requested to base its report on ICRP *Publications 60* and *73*, giving special emphasis to recent accidents and to the lessons to be drawn from them. The scope of the report to be produced was outlined in some detail, and it was requested that recommendations and check-lists be developed.

The membership of the Task Group was as follows:

P. Ortiz López (Chairman)	P. Andreo	J.-M. Cosset
A. Dutreix	T. Landberg	

Corresponding members were:

L.V. Pinillos-Ashton	W. Yin

The Commission is grateful to P.J. Biggs for kindly reviewing the manuscript by correspondence. Numerous helpful comments were received from interested parties through the Commission's customary public consultation via the Internet.

The membership of Committee 3 during the period of preparation of this report was:

F.A. Mettler, Jr. (Chairman)	J.-M. Cosset	M.J. Guiberteau
L.K. Harding (Secretary)	J. Liniecki (Vice-Chairman)	S. Mattsson
H. Nakamura	P. Ortiz López	L.V. Pinillos-Ashton

M.M. Rehani	H. Ringertz	M. Rosenstein
Y. Sasaki	C. Sharp	W. Yin
W.Y. Ussov		

This report aims to serve the purposes described above. In order to be as useful as possible for those purposes, its style differs in a few respects from the usual style of the Commission's publications in the *Annals of the ICRP*. Some colour figures are also included in order to illustrate more clearly the possible consequences of accidents to patients undergoing radiotherapy.

The report was approved for publication by the Commission in October 2000.

Annals of the ICRP

Prevention of accidental exposures to patients undergoing radiation therapy

ICRP Publication 86

Approved by the Commission in October 2000

Abstract-This publication aims to assist in the prevention of accidental exposures involving patients undergoing treatment from external beam or solid brachytherapy sources. It does not directly deal with therapy involving unsealed sources. The document is addressed to a diverse audience of professionals directly involved in radiotherapy procedures, hospital administrators, and health and regulatory authorities. The approach adopted is to describe illustrative severe accidents, discuss the causes of these events and contributory factors, summarise the sometimes devastating consequences of these events, and provide recommendations on the prevention of such events. The measures discussed include institutional arrangements, staff training, quality assurance programmes, adequate supervision, clear definition of responsibilities, and prompt reporting.

In many of the accidental exposures described in this report, a single cause cannot be identified. Usually, there was a combination of factors contributing to the accident, e.g., deficient staff training, lack of independent checks, lack of quality control procedures, and absence of overall supervision. Such combinations often point to an overall deficiency in management, allowing patient treatment in the absence of a comprehensive quality assurance programme. Factors common to many accidents are identified and discussed in detail. The use of radiation therapy in the treatment of cancer patients has grown considerably and is likely to continue to increase. Major accidents are rare, but are likely to continue to happen unless awareness is increased. In this report, explicit recommendations on measures to prevent radiotherapy accidents are given with respect to regulations, education, and quality assurance.

Keywords: Radiotherapy, Accidents, Safety, Radiation protection, QA

MAIN POINTS

- Doses received during radiotherapy are on the upper edge of tolerable doses to normal tissues. As a result, accidental overdosages have often had devastating and sometimes fatal consequences.
- Accidental exposures involving a 10% or more overdosage should be detectable by a well-trained clinician, based upon an unusually high incidence of adverse patient reactions.
- Underdosage accidents are difficult to detect clinically and may only be manifest as poor tumour control.
- Radiotherapy is increasing worldwide and accidents may be expected to increase in frequency, if measures for prevention are not taken.
- While a number of serious and fatal radiotherapy accidents are reported, it is likely that many more have occurred but were either not recognised or reported to regulatory authorities or published in the literature.
- The complex equipment and techniques used in radiotherapy mandate that for accident prevention, there must be sound and risk-informed regulations, managerial commitment at the hospital level, an adequate number of trained staff, adequate resources, a functional implemented quality assurance programme, good communication, and continuing education.
- There is a danger in not fully appreciating that modern equipment and new technologies require more quality assurance and highly qualified maintenance.
- Persons in charge of radiotherapy facilities should ensure that there is proper commissioning of new equipment and proper decommissioning of old equipment and sources.

SUMMARY

(a) Curative or palliative radiation therapy has three major concerns: efficacy, quality of life, and safety (Horiot, 1995). From the point of view of radiation safety, radiotherapy is a very special application of radiation because:

- human beings are directly placed in a very intense radiation beam (external beam therapy), or radiation sources are placed in direct contact with tissue (brachytherapy), to deliver intentionally very high doses (20 Gy–80 Gy), and
- overdosage as well as underdosage may have severe consequences.

(b) From the prescription to the delivery of a radiotherapy treatment, a team of professionals from a number of disciplines is involved in a large number of steps. In the case of external beam radiotherapy, a treatment is usually delivered with 20 to 40 fractions or sessions, each requiring a large number of machine and patient parameters to be set up by the radiotherapy technologist. Based on these considerations, it becomes apparent that there is a significant potential for mistakes leading to an accidental exposure in radiotherapy, and that special preventive measures have to be implemented.

(c) This document aims to assist in the prevention of accidental exposures in radiotherapy, mainly those involving patients undergoing treatment from external beam or solid brachytherapy sources. The document does not directly deal with therapy involving unsealed sources (nuclear medicine radiophamaceuticals). The approach adopted is to: (i) describe illustrative severe accidental exposures that have occurred in radiotherapy procedures during the last 25 years, using case histories[1]; (ii) discuss why these events occurred, highlighting their causes and any contributory factors; (iii) summarise the consequences of these events, which were in some cases devastating; and (iv) provide recommendations on measures to prevent such events, primarily through institutional arrangements, staff training, quality assurance programmes, adequate supervision, clear definition of responsibilities, and prompt reporting.

(d) The document is addressed to professionals directly involved in radiotherapy procedures (radiation oncologists/radiotherapists[2], medical physicists, radiotherapy technologists[3], dosimetrists, radiotherapy nurses, maintenance engineers, etc.) as well as to hospital administrators and health and regulatory authorities. It is intended to provide information and recommendations useful to such a diverse audience; compromises in the scientific detail are, therefore, unavoidable.

[1] It should be noted that, in addition to the accidents reported in detail in this document, a considerable number of other events have occurred in radiotherapy where information could enhance the body of knowledge on this topic. Unfortunately most of these events have not been reported. Not all reported incidents have been included (cf. Järvinen, 1997).

[2] The specialist physician concerned is usually referred to as a radiation oncologist in the United States and as a radiotherapist in Europe. In this report, the term radiation oncologist will be used.

[3] In some countries radiotherapy technologists are called radiation therapists, radiographers, radiotherapy technicians, etc.

Table 1a. Major reported accidental exposures involving patients undergoing external beam radiotherapy and brachytherapy treatments

Country	Year	No. of patients affected	Causes and main contributing factors
USA	1974–76	426	^{60}Co dose calculations based on erroneous decay curve (varying overdoses) No independent verification of dose calculations More than two years without beam measurements
Canada and USA	1985–87	3 deaths from radiation	Accelerator software for the control of treatment features was transferred from other equipment without sufficient consideration to safety
Germany	1986–87	86	^{60}Co dose calculations based on erroneous dose tables (varying overdoses) No independent determination of the dose rate
UK	1988	207	Error in the calibration of a ^{60}Co therapy unit (25% overdose) No independent calibration of the beam
UK	1988–89	22	Error in the identification of ^{137}Cs brachytherapy sources (dosimetry errors between-20% and +10%) No independent determination of source strength
USA	1987–88	33	A computer file (of a treatment planning system) for the use of trimmers was not updated for a new ^{60}Co source because it was not intended for further use The computer file was used (overdoses of 75%) There was no manual verification of the calculated treatment time There was no independent verification of the treatment time
Spain	1990	27 (15 deaths directly from radiation; two deaths with radiation as major contribution)	Error in the maintenance of a clinical linear accelerator; procedures for transferring machine from/to maintenance (notification of physicists) not followed Conflicting signals and displays not analysed Procedures for periodic beam verifications not implemented or insufficient Overdosage ranging from 200% to 600%
UK	1982–91	1,045 (492 patients developed a local recurrence, possibly as a result of the underdosage)	Inappropriate commissioning of a computerised radiotherapy Treatment Planning System (5-30% underdosage) Lack of quality control
USA	1992	1 (death from radiation)	High dose rate brachytherapy source left inside the patient. Source dislodged from equipment Conflicting monitor signals (display at the machine and area monitor) ignored A survey with an available portable instrument was not conducted
Costa Rica	1996	114 (17 deaths from radiation)	Error in the calibration of a ^{60}Co therapy unit. Lack of independent beam calibration and quality control. Recommendations of external audit ignored Overdosages of about 60%

Table 1b. Major accidents due to abandoned radiotherapy sources involving public exposure and/or large-scale contamination

Country	Year	Consequences on humans	Other consequences
Mexico	1984	4,000 people exposed (five of them from 2 to 7 Gy)	30,000 table stands and 6,000 tons of reinforcing rods contaminated 814 buildings partly or totally demolished
Brazil	1988	Four persons died due to radiation exposure 112 persons monitored for contamination	3,500 m^3 of radioactive waste generated
Georgia	1994	One person died due to radiation exposure	
Turkey	1999	11 persons severely exposed to radiation	
Thailand	2000	3 persons died due to radiation exposure	

In most of the accidents in Table 1b the sources were waiting long for a disposal or return to the manufacturer. A major reason for the long delay in the decision was the lack of a disposal capability, the high cost of disposal, and the lack of provisions for the end of life of the sources prior to their purchasing and use.

(e) A summary of the major accidental exposures involving patients undergoing external beam radiotherapy and brachytherapy treatments is given in Table 1a, which also summarises the main factors that contributed to each event. Detailed case histories are given in Chapter 2. Factors contributing to these and other accidental exposures, classified according to the stage in the radiotherapy process at which they occurred, are in Chapter 4. Given the devastating consequences of accidents involving radiotherapy sources, for exposure of the public and/or large scale contamination these are also illustrated in Table 1b (case histories from some of them are in Appendix C).

Causes of accidental exposures in radiotherapy

(f) In many of the accidental exposures reported in this document a single cause cannot be identified. Usually there was a combination of factors contributing to the accident, e.g., deficient staff training, lack of independent checks, lack of quality control procedures, and/or absence of overall supervision. Such combinations often point to an overall deficiency in management, which allowed patient treatment in the absence of essential institutional arrangements, i.e. a comprehensive quality assurance programme. To prevent these events, appropriate institutional arrangements should be required by regulation, and compliance with them should be verified (Ortiz, 1998). The following factors, common to many accidental exposures, can be identified.

Deficiencies in education and training

(g) The lack of accredited professionals with adequate education and training, mainly in the field of medical physics, is a major factor in accidents caused by the erroneous calibration of radiotherapy beams and brachytherapy sources. As illustrated in Table 1a, this type of accidental exposures has involved large numbers of patients. Insufficiently trained radiation oncologists, working without supervision, caused misinterpretations and consequent incorrect treatment. A lack of trained brachytherapy nurses contributed to accidents with radiation sources in the brachytherapy ward. Lack of understanding of beam parameters by a maintenance engineer led to a maladjustment of accelerator energy, causing an accident with devastating consequences to many patients. In most cases, the training of professionals had included only the 'normal' circumstances likely to occur in radiotherapy, but did not consider the identification of 'unusual' situations. As a result, initiating events were not detected and the accidental exposure occurred.

Deficiencies in procedures and protocols

(h) Severe accidental exposures have been caused by inadequate commissioning procedures for radiotherapy equipment (testing and measurement of operational parameters prior to use), computers, and accessories, or by the lack of periodic verification. These accidental exposures could have been avoided if widely available quality control recommendations and protocols had been followed. The use of brachytherapy sources without verification of their activity, or a misunderstanding of the units used to specify their source strength, also caused accidents. Procedures for dealing with equipment repair and maintenance were not followed, and equipment was used for treatment without a previous verification. Deficiencies of procedures and documentation of treatment prescription and planning, patient identification, and patient set-up have also contributed to such events.

Equipment faults

(i) This document does not address issues related to the design of radiotherapy equipment by providing specific recommendations to manufacturers. However, there have been some major accidental exposures, with fatal consequences, as a result of inappropriate design and factory testing that could not be detected by hospital quality controls. In one case, factory tests of accelerator software that had been transferred from an old design to a new one failed to anticipate operating conditions that could occur in practice the manufacturer needed a considerable time to identify the cause, to inform users, and to correct the problem. In another case, equipment faults that appeared sporadically, and could not be resolved by maintenance engineers, led to a decision to operate the machine with interlocks disabled ('physical' mode as opposed to 'clinical' mode); the result was a fatal accidental exposure.

Deficient communication and transfer of essential information

(j) A frequent cause of accidents was misunderstanding of the treatment prescription, treatment plan, or patient identification data. Changes in personnel, without transfer of information on calibration procedures and treatment planning, caused accidental exposures. Deficient communication of patient reactions, conflicting signals and displays, as well as of equipment malfunction contributed to missing opportunities of early detection of accidental exposures.

Lack of independent checks (defence in depth[4])

(k) In many accidental exposures unidentified single mistakes remained undiscovered. These could have been prevented if independent checks had been incorporated into procedures. For instance, mistakes in beam calibration after source installation or exchange could have been identified by independent calibration by a second person or by a consistency check with the manufacturer's source certificate. A systematic application of multiple checks would have drastically reduced accident potential.

Inattention and unawareness

(l) Examples of unawareness included cases when it was not realised that after source change (higher activity), calculated treatment times should be substantially shorter than before. An error in calibration, treatment planning, or calculation should become apparent to attentive staff before treatment delivery. In external beam therapy, treatment delivery involves a large number of steps, utilising substantial data in a repetitive manner, but varying from patient to patient. Radiotherapy technologists often have to enter machine parameters onto treatment charts for upwards of one hundred beams per day[5]. Inattention can lead to the use of a wrong prescription, the wrong patient chart or the selection or recording of wrong parameters. Preparation and delivery of treatment by a radiotherapy technologist requires a continuous and high level of concentration, which may be difficult to maintain when the workload is high. Measures to maintain concentration and awareness require a working environment where distractions are minimised and redundancy checks undertaken (i.e., two individuals should be involved in patient set-up).

Unsecured long-term storage or abandonment of radiotherapy sources

(m) Several major accidents, involving members of the public and with severe environmental consequences, were related to the long-term unsecured storage of radiotherapy sources, either before or after decommissioning.

Summary of recommendations

(n) To prevent accidental exposures in radiotherapy, the following measures are recommended:

[4] The application of more than a single protective measure for a given safety objective such that the objective is achieved even if one of the protective measures fails, for example, multiple independent checks.

[5] This task is achieved by utilising machines with 'record-and-verify' systems.

- regulations for radiation safety should be in place and enforced, and compliance should be monitored;
- appropriate education of professionals directly involved in radiotherapy procedures should be undertaken as a high priority, with continuing education on new treatment techniques;
- a written, comprehensive, and functional quality assurance programme should be adopted and implemented by every radiotherapy institution. This should include a clear chain of responsibilities and procedures for purchasing, acceptance test and commissioning, calibration, treatment planning and delivery, maintenance, two-way communication, and incident reporting.

The future

The benefits of radiation therapy in the treatment of cancer patients are widely acknowledged and as a result, the use of this treatment modality has grown considerably. Usage is likely to continue to increase. Major accidental exposures are rare, but it is likely that they will continue to happen unless awareness is increased. Accidents will usually occur as the result of inadequate education and training, lack of quality assurance, poor infrastructure, equipment failure, and improper decommissioning. Unless these issues are properly addressed and dealt with, more accidental exposures are likely to occur, as current and new technology developments are disseminated.

1. INTRODUCTION

(1) Radiotherapy is concerned primarily with tumour cure or palliation. Modern radiotherapy has three major concerns: efficacy, quality of life, and safety (Horiot, 1995). It is always necessary to be aware of the potential for an accident, the relative importance of human factors, and the wider consequences of an accident.

(2) A radiation accident is an unintended event (operator error, equipment failure, or other mishap) that has or may have adverse consequences (IAEA, 1996). Depending on the type of persons exposed, accidents can be divided into three major groups:

- (i) members of the general public irradiated as a result of failure of implementation of radiation protection and safety rules;
- (ii) clinical staff irradiated during preparation of radiation sources or patient treatment and maintenance staff irradiated during installation, repairs, source change, or other equipment servicing;
- (iii) patients injured during treatment.

(3) The use of the term 'radiation accident' for events involving the public and staff, i.e. groups (i) and (ii), has a relatively straightforward interpretation as neither workers nor public are intentionally irradiated; this makes it easier to distinguish a normal exposure from an accidental one. For patients undergoing treatment, i.e. group (iii), the use of the term 'accident' deserves special consideration. The major focus of this report is in preventing injury to patients. The majority of radiation therapy accidents have occurred in category (iii).

1.1. The potential for an accidental exposure involving radiotherapy patients

(4) The following aspects are relevant to understanding the potential for accidents involving patients:

- Very high doses (20 Gy to 80 Gy) are intentionally delivered to the *'clinical target volume'* (ICRU 1999). As radiation passes through the body, some normal tissues receive doses often similar to that delivered to the target volume, thus resulting in an expected incidence of side effects (Chapter 3).
- The radiation beam is focused directly at the patient (external beam radiotherapy) or sealed radioactive sources are placed in contact with tissues or organs (brachytherapy). Any significant mistake in beam delivery or in the placement of sources can have negative, and often serious, consequences.
- Radiotherapy involves many steps between prescription and dose delivery. Each step may involve a large number of parameters that must be selected, adjusted, recorded, and communicated between different professionals. For example, the delivery of 30 fractions by external beam, each with four fields, requires around 15 parameters to be set for the first field and half this number changed for the other fields; in total, the requirement is to set about 1,000

parameters for the entire treatment. The set-up for each patient is similar, but not identical. The number of parameters is much larger in conformal therapy[6] using multi-leaf collimators and intensity modulated beams, although these are usually computer controlled.

- For treatment units without computerised 'record-and-verify' systems, the radiation technologist has to enter manually parameters in patients' treatment charts. This may be required for up to one hundred treatment beams daily. This is done in a repetitive way, but is different for each patient.
- Sophisticated technology, computer calculations, and data transfer are combined with manual activities e.g. preparation of organ shielding, and immobilisation devices.

(5) The combination of these diverse aspects in radiotherapy requires special and specific safety measures. The design of a safety plan has to consider likely accidental exposure scenarios. Two main approaches are available for identifying these scenarios: retrospective analysis of past adverse events, and prospective methods (event trees and fault trees combined with probabilistic assessment) that may identify additional weaknesses. Situations that have not yet produced an accidental exposure may be overlooked by retrospective methods. However, there are difficulties with prospective methods, e.g. dealing with situations where dissimilar staff training may play a significant role. Both methods have their merits and are complementary.

(6) In this document, retrospective methods are used to analyse several accidents, which occurred in radiotherapy in the period 1974-2000. Data has been obtained from papers published on individual accidents, as well as from regular reports such as those issued regularly by the US Nuclear Regulatory Commission (NRC) and a safety report of the International Atomic Energy Agency (IAEA, 2000a). The number of accidents in Table 3 (Chapter 4) is large enough to identify the major initiators and contributing factors and, in most cases, to provide information on their frequency and consequences. Further weaknesses may be revealed, and additional scenarios anticipated, by a complementary prospective assessment (ICRP, 1997).

(7) The information provided in this report includes primarily classical equipment and treatment modalities. Currently, there is insufficient information about possible accidents involving new technologies and treatment techniques, such as 'gamma knife' units,[7] multi-leaf collimators, intensity beam modulation, intravascular brachytherapy, and high dose rate brachytherapy. Although accidents have already been reported for those new modalities, only prospective methods of analysis may help to prevent accidents.

---

[6] In conformal therapy the cross section of the radiation beam is shaped (conformed) to the target volume, normally using multi-leaf collimators, rather than rectangular collimators. Conformal therapy allows a more efficient and effective treatment, by delivering high-dose to the target volume while keeping doses low to tissues and organs outside the target area.

[7] Several accidents involving gamma knives have been reported recently in the US Nuclear Regulatory Commission Reports to the Congress on Abnormal Occurrences, NUREG 0090, Volumes 17 and 22. One of the cases related to a failure of the patient's couch to retract from the treatment psotion and the couch had to be retracted manually; two other cases related to errors in the preparation of the treatment plan from the prescription and in its implementation.

1.2. Criteria for accidental exposures with radiotherapy patients

(8) In radiotherapy, a 'normal' radiation exposure is a treatment that closely follows the plan specified in the treatment prescription. An accidental exposure can therefore be considered to have occured if there is a substantial deviation from the prescription.[8]

(9) Broadly, the prescription includes: identification of the patient; definition of the target volume; the dose to be delivered to the anatomical site(s); the dose fractionation or treatment schedule; and the final dose distribution. Thus, a departure from the treatment prescription could be: the wrong patient, the wrong anatomical site, a substantial difference from the prescribed dose,[9] the wrong dose distribution (including incorrect beam modality or energy), or the wrong fractionation.

(10) The term 'substantial difference from the prescription' implies a quantification of the difference and its impact on the outcome, i.e., the clinical consequences. Although there are no international standards that include the quantification of a substantial difference or a threshold, there are approaches to classify hazards related to radiotherapy patients. The proposal of the American Association of Physicists in Medicine Task Group 35 (AAPM, 1993) deserves special mention, despite being focused on accelerators and the assumption that weekly quality controls are always performed which would discover errors or equipment malfunctions within one week. This AAPM proposal is based on the US Food and Drug Administration (FDA) regulations, which define a Class I hazard as a condition that could cause death or serious injury and a Class II hazard, where the risk of serious injury is small.

(11) Based on this classification, AAPM (1993) developed a sub-classification of Class I hazards that is valid for the conditions described above. According to this, type A hazards can be directly responsible for complications threatening patient life; type B hazards are those that increase the probability of an unacceptable outcome of the treatment (increased rate of not life-threatening complications or lack of tumour control), but usually do not pose a threat to life. These criteria are summarised in Table 2. Type A hazards are associated with an overdose equivalent to 25% or more of the total prescribed dose. This includes the situation where a dose is delivered outside the intended radiation field, since such circumstances may produce severe clinical complications when a critical organ (e.g., the spinal cord) is near the radiation field. A Type A hazard also includes selection of incorrect beam energy or modality, e.g., irradiation with an electron beam when a photon beam was intended or vice versa. A type B hazard corresponds to overdoses in the range 5%-25% and to most underdosage situations. The interested reader is referred to the original report (AAPM, 1993), where illustrative examples are given.

(12) It is emphasised that inclusion of most cases of underdose in the type B group of events is based on the assumption that the error would be discovered within one week (with weekly quality controls) and corrective actions would be successfully

[8] Situations resulting from an incorrect prescription are not the subject of this report.

[9] For simplicity, in this document the term 'prescribed dose' refers to the dose at different points of the target volume and other tissues and organs, resulting from a clinically accepted dose distribution.

Table 2. Summary of the AAPM TG-35 sub-classification of Class I hazards in radiotherapy (AAPM, 1993)[a]. Remarks added here.

Type	Criteria	Remarks
Type A	25% overdose or more of the total prescribed dose	'The rationale for this choice is related to the observation that a 25% to 50% increase in total dose will often place the patient in the range of the LD50/5 (the probability of 50% lethal complications within five years)...'. For a typical treatment of 40-60 Gy, an overdose of 25% of the prescribed total dose corresponds to 10-15 Gy. This excess in dose can be reached either with an error on each fraction for several fractions during the week or with a large error in a single fraction.
Type B	5% [(i)] to 25% dose excess over the total dose[(ii)]	(i) The value 5% is derived from the TG 35 criteria where an overdosage of 20% during one week corresponds approximately to an overdosage of about 5% over the whole treatment.
	and most underdose situations	(ii) If the underdosage is not discovered within a time in which correction to the treatment can be successfully applied, the hazard should be considered as type A with similar percentage as for an overdose as indicated in the text

[a] Class I hazards are defined by the USA FDA as a condition that could cause death or serious injury. TG-35 considers type A hazards as those that can likely be responsible for life-threatening complications. Type B hazards increase the probability of an unacceptable treatment outcome (complications or lack of tumour control). The criteria refer to a typical treatment prescription of 40–60 Gy total dose with 2 Gy per fraction, and is based on the assumption that weekly quality controls are performed that will discover errors or equipment malfunctions within one week.

implemented during the rest of the treatment. This assumption, however, is not always met. For an example, see case history number 2 of Chapter 2, where the underdosage accident involved 1,045 patients during a 10 year period. Such an accident cannot be considered as being type B. If the underdosage is not discovered within a time period such that corrective actions can be successfully applied, the consequences may be fatal, since the progression of the disease reaches advanced stages and opportunities for remedial actions are not available. Severe underdosage may thus be more adequately classified using limit values similar to overdosages, i.e. an underdose of 25% or more of the total prescribed dose is a type A hazard.

1.3. Clinical detectability of differences in dose prescription and delivery

(13) The question of how large a dose prescription-delivery difference can be detected by observing the clinical outcome of a treatment has relevance in relation to the consequences of human errors or equipment malfunctions. A review of the information on clinically detected dose differences for under- and overdosage situations is therefore relevant:

(i) Dutreix (1984) has reported that a radiation oncologist observed skin reactions on skin folds and also diarrhoea, which were more severe than usual, in patients irradiated for gynaecological tumours with a prescribed tumour dose of 50 Gy to the whole pelvis, given in five fractions per week for five weeks. The effects observed were caused by an error in the calibration of a treatment unit that led to a systematic overdosage to patients. Estimated overdosages were 7% in 67 patients and 10% in 21 patients.

(ii) In a trial study of skin reactions in 58 breast cancer patients, where both internal mammary lymphatic chains were irradiated with electrons, using upper and lower fields with total doses of 50–60 Gy, Wambersie (1984) reported that electron and ^{60}Co adjacent fields dose differences of 10% were detected in 80% of the cases. For symmetric supraclavicular fields, even dose differences of 5% could be clinically detected.

(iii) In a retrospective study of 314 patients with laryngeal cancer, Herring et al. (1971) found a reduction of 20% in the probability of recurrence of the primary tumour in three groups of T3 staged patients treated with doses increased by steps of about 4.5%. The doses administered were 52.5 Gy to the 'low dose' group, 55 Gy to the 'middle dose' group, and 57.5 Gy to the 'high dose' group. An unintended increase in dose, less than 10% above the optimal dose, produced a readily observable increase in the incidence and severity of complications. The increase in dose was due to the omission of a conversion from roentgens to rads that, depending on the accelerator energy, resulted in overdoses between 7 and 10%.

(iv) Differences in local tumour control and survival for groups of patients that differed only by 5% in the average dose delivered have been reported by Dische et al. (1993). In two review studies of the clinical basis for dosimetry accuracy in radiotherapy Wambersie et al. (1994) and Mijnheer (1996) concluded that differences in absorbed dose in the range of 5-10% could be detected in a number of situations.

(14) Although the impact of differences in dose on the probability of tumour control and normal tissue complications depends on the type of tissue irradiated, the fractionation used, the volume irradiated, and other factors (see Chapter 3), from the literature described above it appears that the lowest dose differences clinically detected were in the order of $\pm$5–10%.

(15) On the other hand, it should be noted that the delivery of absorbed dose has an associated uncertainty. The estimated combined standard uncertainty in good radiotherapy practice is in the order of 5% (see Appendix A), i.e., it is of the same order as the lowest dose difference clinically detectable. The larger the size of the error, the larger the probability of the undesired outcome and, in the case of normal tissue complications, the greater the severity. When the error size is large enough, the unplanned detrimental outcome becomes virtually certain. This was the case in some of the accidental exposures described in Chapter 2.

(16) Traditionally, only patient overdose has been considered of importance. However, patient underdosage is as important as overdosage, and in certain instances

it can be even more significant (see the case described in Section 2.2 and Chapter 3). Unfortunately, cases of underdosage are usually more difficult to detect clinically than overdosages, and they are usually only discovered long after the treatment has been completed and there is a higher than expected incidence of tumour recurrence.

1.4. Other events of interest to the prevention of accidental exposures

(17) There is a group of events that can be considered accident 'precursors' (i.e., do not have clinical consequences but have the potential to develop into accidents with such consequences). For example, events detected by 'good luck', or by a quality assurance programme including redundant checks and systems, could result in an accidental exposure if the same events occur in another institution.

(18) The compilation of information on such events, their evaluation to determine generic implications and lessons that can be learned, and the dissemination of this information is highly desirable to prevent accidental exposures. Some national systems for reporting, such as that in the USA, include provisions to report 'events of interest' that are considered precursors, even if they do not involve a major reduction of the level of protection. Similar programs are being developed by professional organisations.

2. CASE HISTORIES OF MAJOR ACCIDENTAL EXPOSURES IN RADIOTHERAPY

(19) This section provides a short description of some major accidental exposures, as an illustration of how initiating events can develop into severe, large scale, accidental exposures with fatal consequences (IAEA, 2000a).

2.1. Incorrect ^{60}Co decay chart and lack of verification (USA, 1974-76)

(20) A cobalt (^{60}Co) teletherapy unit had been calibrated correctly initially, but subsequent dose calculations were based on an incorrect ^{60}Co decay curve. No periodic measurements of the machine output were made. The decay curve had been drawn on a linear scale with an incorrect slope, resulting in a calculated decay more rapid than that of the real source. The dose rate was therefore underestimated and consequently longer treatment times were calculated from the erroneous decay curve.

(21) Initially, the incorrect machine output was attributed to a faulty measuring system. The medical physicist produced ten calibration documents showing that the machine output used was correct. When some patients showed symptoms of overexposure, the hospital asked an external consultant physicists to review the dosimetry. They found that the ^{60}Co unit and measuring system were functioning correctly, but the resident medical physicist had fabricated all but one of the reports on machine output.

(22) Overdoses to patients had increased progressively over 22 months. The level of overdose was 10% in the first 5 months and increased to as high as 50% by the end of the period. During the last 16 months, 426 patients were treated.

(23) Approximately 20 years after the accidental exposure, Cohen et al. (1995) evaluated the clinical consequences using the medical records of the 26 patients. Three hundred patients died in less than one year, mainly due to their cancer. Of the patients who survived the first year, 88 had immediate severe complications related to the irradiated sites (including brain, spinal cord, skin, oropharyngeal mucosa, colon, and rectum). The clinical follow-up notes pertaining to the period between one and three years after the treatment showed that a high proportion of patients developed significant, often life-threatening, complications. These included severe skin reactions with ulceration, mucosal reactions with necrosis, stenosis of pharynx and esophagus, ulceration and/or perforation of stomach and bowel, bone necrosis, and myelopathy.

(24) A contributory factor was insufficient staff resources to assure adequate dosimetry and quality control. The actual beam output was not checked over a 22-month period, and the medical physicist was dedicated to supporting a new accelerator; there was no independent check of the decay curves; the physicist falsified records afterwards in an attempt to justify his calculations.

2.2. Lack of acceptance procedures for a treatment planning system (UK, 1982-1990)

(25) Until 1982, manual calculations were the only method available for treatment planning in a specific hospital. Treatments were usually performed with a fixed SSD

(source-skin-distance) technique. For treatments requiring an SSD other than the usual 100 cm, a procedure had been established based on correction factors for varying distances that radiotherapy technologists applied manually.

(26) A computerised treatment planning system was acquired in the autumn of 1982. This allowed isocentric treatments on a more frequent basis. Since isocentric techniques are carried out with distances shorter than the usual 100 cm SSD, for the first isocentric treatment the technologists (with the acquiescence of the medical physicist) used the distance correction factors previously established. Every subsequent isocentric treatment had this manual calculation for the distance correction applied.

(27) It was not realised that the computer planning software already incorporated corrections for varying source-skin distance. The correction was therefore duplicated when the radiotherapy technologists continued to manually apply distance correction for distances other than 100 cm. Since no quality control measurements were made, the error was not detected. Thereafter, the practice of manually applying the distance correction factors every time that an isocentric treatment was used remained. At the end of 1991, almost a decade later, the error was discovered.

(28) The error caused delivered radiation doses to be lower than the prescriptions in all isocentric treatments during that period of time, which involved 1,045 patients. A study made by Ash and Bates (1994) on these patients concluded that 492 patients developed a local recurrence, possibly as a result of the underdosage. For 189 patients it was not possible to be certain about the effects of the underdosage. For 320 patients, it was considered unlikely that they had suffered any effects from the underdosage, and in the remaining 44 patients the outcome was unknown.

(29) An independent inquiry into the case found the following factors contributing to this accident: the responsibility and authority of the medical physicist were not clearly defined; the number of staff members was insufficient for the number of patients treated and, in particular, the shortage of medical physicists led to overloaded working conditions; a systematic quality assurance programme that included commissioning following the introduction of new equipment was not in place; and the correction procedure followed for nearly a decade was never embodied in writing.

2.3. Six accidental exposures involving software problems in several accelerators of the same type (Canada and USA, 1985-87)

(30) Between June 1985 and January 1987, six known accidental exposures involved massive overdose by the same type of accelerator (Leveson, 1993). Several flaws were identified in the software used to enter selected parameters for the treatment such as type of radiation and energy. Two of these flaws caused the deaths of three patients.

(31) The first accident leading to a fatality occurred in March 1986. After selection of the treatment parameters for x-ray mode, the operator changed the selection from x-ray to electron mode. During this time the equipment was executing the initial request, which required the automatic rotation of a turntable assembly with the bending magnet for beam scanning (electron mode) and the target and beam

flattener (x-ray mode). The equipment did not execute the second request correctly and the turntable had not reached the correct position. The equipment then operated without the x-ray target and the beam flattener, thus leading to massive overdoses.

(32) The outcome was radiation-induced myelitis of the cervical cord causing paralysis of the patient's left arm and both legs, left vocal cord paralysis (which left him unable to speak), neurogenic bowel and bladder, and paralysis of the left diaphragm. He died from complications of the overdose five months after the accidental exposure.

(33) The maintenance engineer was unable to reproduce the malfunction and the problem remained unresolved. The same problem reappeared in April 1986 in the same hospital and a second patient was overexposed. He developed disorientation that progressed to coma and neurological damage from high-dose radiation injury to the right temporal lobe of the brain and the brain stem. The patient died three weeks after the accident.

(34) The medical physicist in the hospital was finally able to reconstruct the fault. This forced the manufacturer to respond (Rawlinson, 1987). The exact doses in these cases are uncertain but they were estimated to be in the order of 150 Gy- 250 Gy in one single treatment fraction (delivered within a few seconds).

(35) A different software problem caused the death of a patient in 1997 in another hospital. An operator pressed the 'set' button at the precise moment when a variable in the software rolled over to zero. This allowed the machine to start irradiation without x-ray target and without scanning of the electrons over the radiation field. The result was a highly concentrated electron beam.

(36) There were a number of problems with the design, testing, and process for following up of equipment faults, as well as in the dissemination of warnings to users of the same type of equipment. It is not possible to summarise these problems in a few lines. Interested readers are referred to the detailed reports of Rawlinson (1987) and Leveson (1993). Two of the major problems were:

- The software package for controlling some of the functions in the accelerator was reused from an older type of accelerator of a significantly different design. In the new accelerator type, safety of these functions relied entirely on the old software. In a review study (Leveson, 1993), it was concluded that using software modules designed for other equipment does not guarantee safety in the new system to which they are transferred and sometime leads to awkward and dangerous designs. Safety is a quality of the system in which the software is used; it is not a quality of the software itself.
- There was no efficient mechanism to follow up reports of suspected accidents. The first accident occurred in June 1985 and resulted in a mastectomy for radiation complications. The patient completely lost the use of her shoulder and arm. In spite of this, the manufacturer and operators of the machine refused to believe that it could have been caused by the accelerator (Leveson, 1993). The accident was not reported to the US Food and Drug Administration (FDA) until 1986. This considerably delayed the investigations by the FDA. The reporting regulations at that time applied only to equipment manufacturers

and importers, not users. The regulations were amended in 1990. Other users of the same type of machine remained unaware of the problem and another accident occurred seven weeks later in Canada.

2.4. Computer file not updated for ^{60}Co source change (USA, 1987-88)

(37) A computer programme file used for the treatment of brain cancer with 'trimmer bars' was not updated in March 1987 when the therapy department replaced the ^{60}Co source. Other programme files had been revised, but this aforementioned file had not, because 'trimmer bars' were not used at that time to treat patients with whole brain irradiation. However, in September 1987, brain treatments were restarted and the 'trimmer bars' computer file with the data corresponding to the prior source was used.

(38) As a consequence of the error, 33 patients undergoing radiation therapy to the brain had received radiation doses that exceeded the prescribed dose by 75%. By the time of the notification of the case to the State Agency, 20 patients had died either during or after the conclusion of treatment (NRC, 1988; 1996).

2.5. Incorrect repair followed by lack of communication (Spain, 1990)

(39) Following instability of the radiation beam of a linear accelerator, the interlock system terminated the treatment of a patient. The problem had arisen at the beginning of a long holiday weekend. An engineer from the maintenance company who had been adjusting an adjacent ^{60}Co therapy unit was requested to 'have a look' at the accelerator. Over the long weekend the maintenance engineer made several unsuccessful attempts to identify the cause of the machine fault. Eventually, he manipulated the energy control system to be able to restore the radiation beam.

(40) The following Monday, patient treatments resumed. There were established procedures for notifying the hospital's maintenance engineer and medical physicists, but these were not followed. If they had, beam tests would have been performed, as required after any repair that could affect the dosimetry of the accelerator.

(41) A control panel meter indicated 36 MeV permanently, regardless of the energy selected; i.e., for an electron energy selection of 10 MeV the selector light button indicated 10 MeV but the control panel meter indicated 36 MeV. The radiotherapy technologists operating the accelerator pointed out the discrepancy between these two indications to the maintenance engineer, but his explanation was that the meter of the control panel was mechanically stuck on 36 MeV. He did not realise that his attempts to restore the radiation beam had forced the beam energy control to be at its maximum. He failed to compare the indication of the control panel meter with any other parameter related to the accelerator energy (e.g., on the modulator in the machine room). He simply assumed that the control panel meter was stuck and that the actual energy was the one indicated by the selector light button.

(42) The reality was the opposite: the control panel meter indicated the actual energy and the energy selection button did not work at all. Treatments, therefore, resumed without any beam measurements. The medical physicists were only notified of the

discrepancy between the two indicators on the 10th day after resuming treatments. They measured the beam output and energy and concluded that both were incorrect.

(43) The consequences of the mistake were relevant only to treatments with electron beams, for which the delivered dose and beam energy (and therefore the electron dose distribution) were substantially different from the selection made at the operating console of the accelerator. The penetration of the electron beams was that of 36 MeV electrons even when smaller penetration depths were intended by selecting low energies; therefore, large doses were delivered to tissues deeper than intended. The dose delivered was also much higher than intended due to the 'focusing' effect of electrons into a smaller cross section of the beam. The dose rate was between 3 and 7 times higher than intended, depending on the selected energy.

(44) The subsequent court proceedings (El Periódico, 1993) established that during the 10 days of malfunction, 27 patients had been treated with electrons; 15 died with radiation as the primary cause, two died with radiation as a major contributor to their death and others suffered major disability. The main radiation effects contributing to death were respiratory insufficiency, bronchopneumonia, radiation myelopathy, oesophageal stenosis and perforation, renal insufficiency, cervical haemorrhage, and arteriosclerosis. Major effects on patients alive at the time of the court decision were paralysis and severe disability.

2.6. Malfunction of brachytherapy high dose rate equipment (USA, 1992)

(45) A patient was to be treated using a high dose rate (HDR) brachytherapy unit equipped with a 16 GBq (4.3 Ci) ^{192}Ir source. The prescribed dose was 18 Gy in three fractions. Five catheters were placed in the tumour and the source was to be stepped through the pre-programmed positions in each catheter. During the first fraction the radiation oncologist experienced difficulties in positioning the source into the fifth catheter and decided to retract the source. The source became detached from the driving mechanism while still inside the patient.

(46) The staff disregarded an alarm from an external area radiation monitor because the console of the brachytherapy unit indicated 'safe'. All three technologists and one physician who were attending the patient were aware of the alarm condition but none of them conducted a survey with the available portable radiation survey instrument (NRC, 1992).

(47) The patient, with the source still in the catheter, was transported back to the nursing home. The source remained inside the patient for almost four days, until the catheter containing the source fell out. The patient received a dose of 16,000 Gy at 1 cm distance from the source, instead of the prescribed 18 Gy. The nursing home staff disposed of the catheter in an area used to store non-radioactive medical waste and was removed later by an incinerator company. The source was discovered when it tripped a radiation monitor located at the incinerator.

(48) The patient died shortly after the source was dislodged. The overexposure was the major contributing cause of death. The lost source also caused radiation exposure to 94 other individuals, including persons at the cancer clinic, nursing home, ambulance staff, and workers at the waste-disposal company. A similar accident

in another hospital was subsequently avoided because the medical physicist was aware of the first case and immediately recognised the problem. This emphasises the importance of incident reporting and dissemination of the lessons learned.

2.7. Beam miscalibration following the exchange of a ^{60}Co source (Costa Rica, 1996)

(49) A hospital had two ^{60}Co teletherapy units. Inconsistencies in dosimetry at the hospital had been detected before the accident by the IAEA/WHO TLD postal dose quality audit service. An external expert was sent who confirmed the deficiencies in dosimetry procedures and the lack of appropriate quality assurance. These findings were reported to the hospital during the audit, but were not acted upon. Insufficient education in medical physics, lack of an independent calibration, the lack of a quality assurance programme, the absence of documented procedures, and the lack of awareness at the hospital, allowed mistakes to remain unnoticed.

(50) After a source exchange in one of the ^{60}Co machines the medical physicist incorrectly calibrated the beam output; using a time of 30 s instead of 0.3 min (18 s) to determine the absorbed dose rate. This resulted in a 66% overestimation of the exposure time and therefore in an underestimation of the dose per unit time, which led to longer treatment times.

(51) The radiotherapy technologists questioned why patient treatment times remained about the same as with the old source when a new source was in place. The medical physicist told them that the delivered doses were correct according to his calculations. The radiation oncologist saw the patients only occasionally and ignored the high incidence of side effects such as vomiting, diarrhoea, and rectal bleeding. Over a period of about one month 114 patients received considerable overdoses, resulting in severe health effects including fatalities. The accident was finally detected by another radiation oncologist at a different hospital who noted severe complications in his patients that had been treated on that machine.

(52) Two assessments of the consequences were performed by an international medical team, one and two years after the accident (IAEA, 1998b). Of 51 patients who died within the two years after the accident, 13 were radiation related and 4 possibly related. Of the 52 patients who where alive two years after the accident, 4 showed 'severe or catastrophic' effects due to radiation overexposure and 12 showed marked effects, with a high risk for future effects. The radiation effects found in patients that survived two years after the accident were:

- Nervous system: Brain-atrophy, necrosis, decreased cognitive function, headaches, mood alteration, seizures, decreased intellectual function. Spinal cord paralysis, quadriplegia, paraplegia.
- Skin: fibrosis, atrophy, contraction, induration, edema, pigmentation, puritis, hypersensitivity, pain.
- Lower gastrointestinal: chronic or bloody diarrhea, bowel stenosis, stricture, fibrosis, obstruction, fistula perforation.
- Bladder: dysuria, hematuria, contracture, incontinence.
- Vascular and lymphatic: stenosis, premature atherosclerosis.

Figure 1 (see frontispiece) shows some of these effects. In addition, there were considerable psychosocial effects (IAEA, 1987).

3. CLINICAL CONSEQUENCES OF ACCIDENTS IN RADIOTHERAPY

3.1. Side effects and complications—normal therapy versus accidental exposures

(53) In radiotherapy, a number of normal tissue side effects or complications are expected and accepted as part of treatment since some normal tissues will receive doses of the same order as the tumour dose. This cannot be avoided in most cases. Fig. 3 illustrates this principle (adapted from Pérez and Brady, 1992). The figure also shows that a departure from the prescribed optimum dose will result in either an increase in the number of complications or in a decrease in tumour control.

(54) Side effects are usually considered to be minor and transient. Complications are more severe and long lasting. Side effects can be accepted at a high frequency while complications will only be acceptable at a lower frequency. Traditionally, tolerance doses in radiotherapy have been set at a '5-5' level, i.e. 5% complications at 5 years (Rubin, 1968). This approach is presently changing, for a number of reasons:

- Firstly, the 'acceptability' of side effects and/or complications actually varies according to their type. In cancer treatments, even a high percentage of relatively minor side effects, such as xerostomia or localised subcutaneous fibrosis, is acceptable. In a survey, a 5% rate of very severe complications (such as

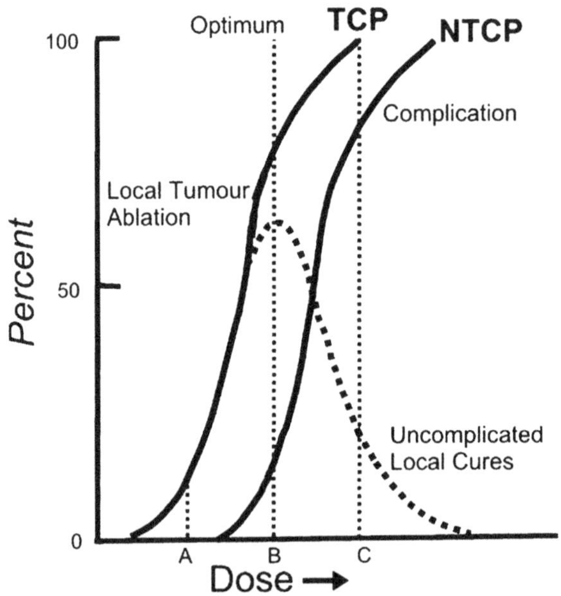

Fig. 3. The solid curves represent the probability of tumour control (TCP) and of normal tissue complication (NTCP) versus delivered dose. The dashed curve represents the probability of uncomplicated tumour control with its maximum at dose B. This dose represents the optimal balance between local tumour ablation and an acceptable incidence of side effects or complications. Dose C could achieve 100% of cure rate but, at this high level, because of complications most patients would not survive the treatment. At dose A there would be no complications, but the probability of controlling the tumour would be very small.

radiation myelitis) was considered as unacceptable by most radiation oncologists in curative radiotherapy (Sebag-Montefiore, 1992).

- Secondly, techniques of radiotherapy are regularly being improved; the recent developments in conformal radiotherapy have allowed radiation oncologists to increase tumour dose to improve cure rate, without concomitant increases in toxicity.

(55) These side effects and/or complications of radiotherapy (which are acceptable in cancer treatment) should be clearly distinguished from the detrimental consequences of an accidental overdosage. The former are an intrinsic part of the prescription, representing a benefit-risk assessment. The latter are unplanned events which have no relation to a benefit-risk analysis.

(56) An accidental exposure in radiotherapy can also be the result of an unplanned and significant underdosage of the target volume. In this case, side effects or complications will be absent or perhaps minor in normal tissues, but the probability of cancer cure may be substantially decreased.

3.2. Consequences of accidental exposures in radiotherapy

(57) The consequences of accidental exposures can be categorised into three types: (i) impact on local tumour control rate, (ii) early (or acute) complications, and (iii) late (or chronic) complications. These consequences will be briefly discussed in the following subsections, concentrating on the physical and biological parameters that affect local tumour control and toxicity.

3.2.1. Impact on tumour control rate

(58) The delivery of a tumour dose significantly *lower* than the prescribed dose, as a result of an accident, can severely jeopardise the probability of curing the patient. Although infrequently reported in the literature, this can lead to the death of the patient (or a group of patients; see case history in Section 2.2) due to disease progression. Such errors are frequently the cause of lack of tumour control. These errors may not be identified for a long time, and consequently may involve a large number of patients.

(59) In the case of an accidental overdosage, the tumour control probability may increase. However, there will be an associated overdose to normal tissues, as described in Sections 3.2.2 and 3.2.3, leading to death or to a severely reduced quality of life. The choice of the prescribed dose is based on an acceptable benefit/risk ratio (see Section 3.1), and any unintended deviation is undesirable.

3.2.2. Early (or acute) complications

(60) Acute complications are early deterministic effects (effects due to cell killing). They are dose related and have a threshold (i.e., below a certain dose they are not seen). These effects are usually observed in tissues or organs with rapid cell turnover

rates (e.g., skin, mucosa, and bone marrow). These complications are observed within days or weeks after irradiation. They are often transient.

(61) The main determinants of early effects are: (1) the *delivered dose*; (2) the *total duration* (protraction) of the radiation treatment (because of the ability of rapidly proliferating normal tissues to compensate for cell loss during the few weeks of a conventional, or accidental, irradiation); and (3) the size and location of the *irradiated volume*- or percentage of the irradiated organ. In contrast, there is a low correlation between early complications and fraction size or dose rate, except when the latter is very high.

(62) In overdosage accidents, the severity of the early side effects or complications is increased. Ultimately, if the overdose is very high, all tissues within the irradiated volume will be destroyed (e.g., radionecrosis will occur with skin doses over 25 Gy in one fraction, particularly if the field is larger than 2 cm diameter).

3.2.3. Late (chronic) complications

(63) Complications of this type are also deterministic effects and have a threshold (see Fig. 3). These late effects are mainly observed in tissues or organs with slowly proliferating cells. They can also be seen in organs with rapidly proliferating cells, a consequence of very severe acute reactions ('consequential' effects; Maciejewski, 1990). The physiopathology of these late complications is complex, a combination of cell loss, alteration of supportive tissues (e.g., capillaries, glial cells), and the progressive and chronic development of radiation fibrosis. These late complications usually occur more than six months after the end of irradiation, but can be observed much later (several years). They are usually considered as irreversible, and are often slowly progressive.

(64) The *delivered dose* is a major determinant of these late effects. The slope of the dose/effect curve is even steeper than for early (acute) effects. *Fractionation* or fraction size (for external beam) or dose rate (for brachytherapy) has been shown, both experimentally and in practice, to have a major impact on late effects, even with relatively modest changes in fraction size or dose rate.

(65) In the case of an accidental increase in fraction dose, this 'fractionation sensitivity' of late effects amplifies the toxicity of an already increased total dose (Thames et al., 1987; Cosset et al., 1994). In some accidents, increased doses per fraction have clearly amplified the severity of the injuries (see Section 2.7). Treatment with only one beam, when the prescription is for several beams every day, results in relatively high fractional doses to some tissues close to the entrance beam. This increases the late effects, similar to using larger doses per fraction. In accidental overdoses, this further increases the severity of complications. Similarly, in low-dose-rate brachytherapy, an accidental increase in dose rate enhances the severity of late complications for the same total dose.

(66) The severity of the observed late complications is closely related to the volume of tissue irradiated above the threshold in organs where subunits are arranged in parallel, such as lung and liver (see Fig. 4). In contrast, in serially arranged organs such as spinal cord, intestinal tract or large arteries, a lesion in a subunit

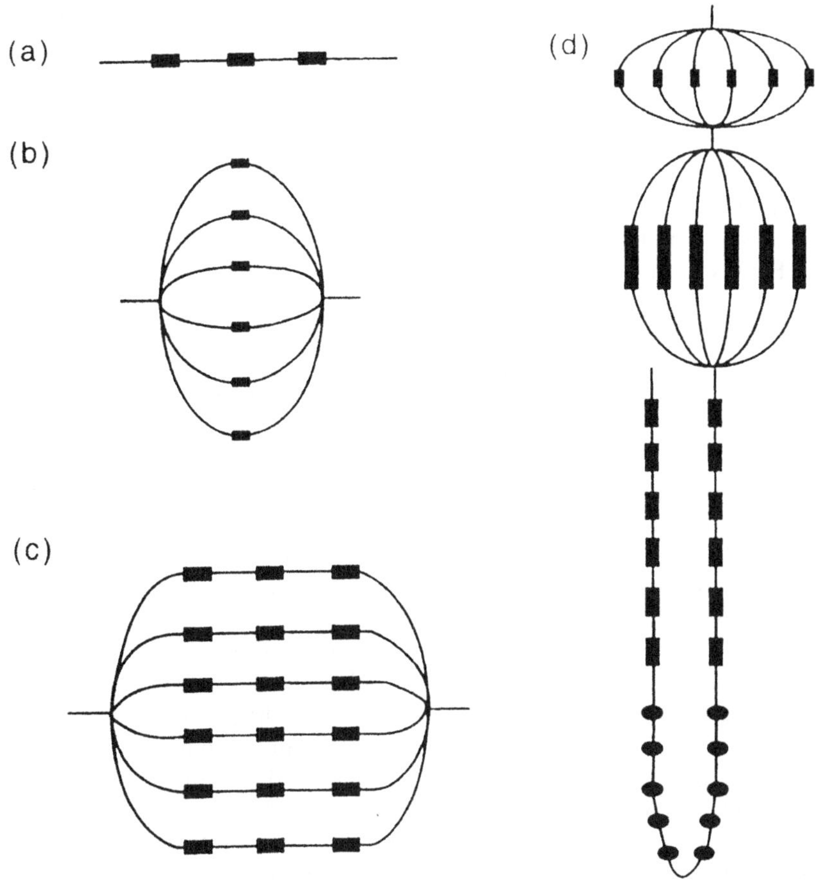

Fig. 4. Schematic examples of tissue organisation structures in the parallel-serial model: (a) a serial string of subunits (e.g., the spinal cord); (b) a parallel string of subunits (e.g., the lungs); (c) a serial-parallel string of subunits (e.g., the heart) (d) a combination of parallel and serial structures (e.g., a nephron). From ICRU (1999).

significantly affects the function of the whole organ (see Fig. 4). In such serially arranged tissues irradiation of a very small volume above the threshold may lead to highly incapacitating complications. Complete paralysis caused by a radiation-induced transverse myelitis is an example of a major complication related to the irradiation of a very small[10] volume in a serially arranged organ. In contrast, total *duration* (protraction) of irradiation has a minor role in late complications (except in the case of 'consequential' late effects, after very severe acute reactions).

[10] While most published work regards spinal cord as a serially arranged tissue, some authors consider that this may not be strictly correct (Powers, 1998). In practice, these considerations should not lead to underestimation of the consequences of irradiation of these organs.

(67) In summary, in overdosage accidents, where the patient survives the early/acute period, late complications will mainly be related to *total dose,* to the *dose per fraction,* to the *type of tissue* and to the *volume* of the organ irradiated. The duration of the radiation treatment (protraction) has only a limited impact. These late complications can lead to very severe detriment for the patient and can be fatal.

(68) In addition to the late (deterministic) complications, secondary malignancies are also late (stochastic) effects of irradiation. These effects are not due to cell-killing but are related to mutations. An increase in dose increases the probability of second cancers. However, the precise relationship between dose and carcinogenic risk in the radiotherapy dose range is still debated. When these secondary radiation-induced malignancies occur their severity is independent of the radiation dose.

3.3. Impact of individual radiosensitivity

(69) In the past, the claim that some individuals were hypersensitive to ionising radiation was frequently used as an explanation for some unexpected side effects or complications. However, it is probable that most such cases were related to accidental overdosages. The development of quality assurance programmes has demonstrated that the majority of unexpected side effects and complications were due to the overdoses (accidents) and not to individual increased sensitivity to radiation. A very small subgroup of cancer patients can be identified for whom individual hypersensitivity may pose a problem during (or after) radiotherapy. However, this can only be established after careful and thorough elimination of the possibility of an overdose. The number of hypersensitive patients in a cancer population may be in the range of 0.5–3%. This percentage may be much higher than that estimated for the normal population who do not have cancer. The reasons for this are: (1) patients with well-known and very rare hypersensitivity syndromes, such as ataxia-telangiectasia, are much more likely to develop a cancer than non-affected individuals; and (2) patients with genetic profiles predisposing to cancer have been related to alterations of DNA repair mechanisms, and consequently they may be hyper-radiosensitive (ICRP, 1998).

(70) Within this small group of cancer patients, two subpopulations can be distinguished. The first one corresponds to the known rare hypersensitive syndromes (essentially ataxia-telangiectasia and Nijmegen breakage syndrome). Since the diagnosis of these syndromes precedes the emergence of a cancer in almost all cases, it should be possible for the radiation oncologist to modify the therapy (Sharp, 1999). In some cases, radiotherapy will be simply avoided, in others, fraction size and total dose will be reduced (by a factor of about 3–4).

(71) The second group of hypersensitive patients, which represent the majority, cannot be clinically distinguished *a priori* from other patients. Currently, an increase in early and/or late reactions is the only way to identify such patients. Recent studies suggest that in some of these patients, alterations of DNA repair mechanisms could be responsible for their enhanced radiosensitivity, but this is not the only factor. To date, it has been difficult to find assays that will consistently predict radiosensitivity in these groups before radiotherapy commences.

3.4. Clinical detection of accidental exposures in radiotherapy

(72) Careful clinical follow up may detect an overdose accident before the total treatment course is completed. Overdoses often cause early/acute, abnormal (enhanced) reactions. These should be promptly detected by an experienced radiation oncologist during regular weekly consultation. As mentioned earlier, dose variations as low as 7-8% have been clinically detected (Dutreix, 1984). When an unexpectedly severe reaction is seen, it is essential to assess all patients treated on the same machine, and possibly in the same unit/department.

(73) Some overdoses will cause late severe effects without abnormal acute (early) effects. Consequently it will be only possible to detect such accidents during long-term follow-up. Regular follow-up is therefore of paramount importance. In the case of unusual reactions in a single patient, it will be useful to recall all other patients treated at the same period of time with the same machine, and possibly in the same unit/department.

(74) In underdose accidents, clinical detection is much more difficult. However, an experienced radiation oncologist should be able to identify a decrease in the normally observed early side effects, especially if this affects several patients at the same time. When the decrease in side effects or complications is insufficient to attract attention, decreased tumour control in a number of patients may raise concern that underdoses have been given. Evaluation of the mid and long term results of radiation treatments should be undertaken and may allow the detection, even if late, of some types of accidents.

(75) When there are very severe consequences of an accident, such as massive and extended radionecrosis, the management is often complex and very difficult, and the results of treatment are frequently disappointing. Massive radiation-induced fibrosis is almost irreversible, and large areas of necrosis may require complex therapies, including reconstructive surgery. This reinforces the requirement to detect an accident at the earliest possible opportunity, in order to minimise complications and fatalities.

4. CAUSES OF AND FACTORS CONTRIBUTING TO ACCIDENTAL EXPOSURES IN RADIOTHERAPY

(77) Event reporting is essential to improving safety. Dissemination of information on radiotherapy accidents and their causes will help to prevent future accidents. This will only be effective if the information has been compiled and analysed. Each accident is triggered by an initiating event. Such events normally do not progress to an accident if a well-designed quality assurance programme is in place. Well-designed quality assurance programmes contain sufficient layers of safety (physical or procedural or both) to prevent progression of initiating events into accidents.

(78) This section provides a short description of initiating events together with a summary of the main contributing factors and omissions. These are used to provide recommendations on accident prevention that are given in Chapter 5. The information is from published papers on individual accidents, as well as from reports issued regularly by the U.S. Nuclear Regulatory Commission (NRC) and a safety report of the International Atomic Energy Agency (IAEA, 2000a).

(79) The information in Sections 4.1 and 4.2 concerning external beam therapy and brachytherapy is classified into the groups described in Table 3, ordered to align with the treatment process. The topic of radiation sensitive patients was discussed in Section 3.3 and will not be revisited below. Public exposures and environmental contamination are discussed briefly in Section 4.3. Section 4.4 summarises generic lessons learned.

4.1. External beam

4.1.1. Equipment problems

(80) There have been software problems leading to equipment faults under certain operating conditions, as described in Section 2.3. Accidents with the same type of accelerator occurred in six different occasions over a period of two years and three patients died. The contributing factors were:

(i) a software package was transferred from an from an older accelerator type and without effective assessment of all safety implications;
(ii) the equipment faults were difficult to reproduce; and

Table 3. Classes and frequencies of accidental exposure in radiotherapy

Accidental exposures in external beam therapy	No. of cases	Percentage of cases (rounded)
Equipment problems	3	6.5
Maintenance	3	6.5
Calibration of the beams	14	30
Treatment planning and dose calculation	13	28
Simulation	4	9
Treatment set-up and delivery	9	20 (**)
Total	46 (*)	100
Accidental exposures in brachytherapy		No. of cases
Equipment and source problems	5	15
Source order and delivery, calibration, and acceptance	3	9
Source storage and preparation for the treatment	5	15
Treatment planning and dose calculation	6	18
Treatment delivery	11	34
Source removal and return	3	9
Total	33 (*)	100

*The number of accidents in the table are fewer in number than in the source publications, since the source publications include events with unsealed sources and accidents involving the public.

**It is likely that errors in the treatment set-up are more frequent than tabulated, since many instances probably remain unreported, especially if the consequences are moderate, i.e., affecting one or a few fractions.

(iii) the difficulties in identifying the cause led to a long delay in disseminating warnings and taking corrective actions

4.1.2. Maintenance

(81) There have been two major accidents related to maintenance problems. Both accidents resulted in deaths. In the first of them, the initiating event was the maladjustment of the electron energy in an accelerator (see the case history in Section 2.5). The second accident was due to intermittent failures, followed by frequent interruption of treatments for repair work and several unsuccessful attempts to repair the machine. This situation eventually led to interlocks being disabled. Contributing factors were:

(i) Insufficient knowledge on the part of the maintenance engineer about the consequences of manipulating beam parameters, such as the energy, and deficiencies in training and expertise in diagnosing the cause of equipment faults.
(ii) Transfer of the machine from the hospital staff to the maintenance engineer and return without communication to the medical physicists about the maintenance. Treatment resumed without dosimetric check of the beam.

(iii) Operation with the energy selector or other key features disabled.
(iv) Conflicting display and signals were ignored.
(v) Lack of equipment for quick constancy checks.
(vi) Intermittent faults, difficult to reproduce, identify, and repair.

4.1.3. Beam calibration

(82) The most important factor in accidental exposure in radiotherapy is an error in the determination of the machine 'output' (dose per time unit). Typically, these occur when there is an error in the beam calibration, or incorrect calculation of decay. Other errors involve misinterpretation of calibration certificates and errors in correction for atmospheric pressure. In one case, data reported by the weather station was corrected to sea level but the user assumed that the data was related to pressure at the level of the weather station. In a different accident, a ^{60}Co beam was used without calibration (the exposure value in Roentgen from the radiation source certificate was simply taken as cGy). In yet another accident, a plane-parallel chamber was used incorrectly (upside down), because the label was misplaced and a new physicist was unfamiliar with the chamber.

(83) Factors contributing to these accidents were:

(i) Insufficient training on beam calibration procedures; insufficient understanding of calibration certificates, dosimetry equipment, and conditions to determine atmospheric correction factors.
(ii) Lack of redundant and independent absorbed dose determination.
(iii) Lack of clear procedures and protocols, and of overall supervision of compliance with procedures.
(iv) Changes of personnel (physicist) with poor communication and transfer of information between staff.

4.1.4 Treatment planning systems

(84) Two major subgroups of accidents can be distinguished in this area: (a) accidents related to the commissioning of a radiotherapy treatment planning system (TPS) or entry of incorrect basic common data, and (b) accidents due to mistakes with individual patients. Two accidents of the (a) type were discussed in Chapter 2. In the case in Section 2.2, the event affected 1,045 patients and was related to operation of the TPS without proper commissioning. In the case in Section 2.4, there was a failure after a source change to enter new data into one of the computer files for certain type of treatment.

(85) Factors contributing to group (a) accidents were:

(i) Insufficient understanding of the treatment planning system, TPS (no training of the staff on the new TPS).
(ii) Lack of formal commissioning of the TPS (no comprehensive operational tests before using the TPS for treating patients).

(iii) Lack of an independent check of the planning (either by manual calculations to selected points or by measurement in a phantom).

(86) Factors contributing to group (b) accidents were:

(i) Lack of understanding of the TPS.
(ii) Lack of independent check of the treatment planning.

4.1.5 Treatment simulation

(87) In one incident the wrong side of the patient was treated due to incorrect labelling of the simulator film. As a result 2 Gy was given to the right side instead of the (intended) left side.

(88) Contributing factors were:

(i) Treatment simulation in an unusual position.
(ii) No check of the anatomical site relative to the film.

4.1.6 Treatment set-up and delivery

(89) In one accidental exposure in this category, a patient responded when another patient was called and a fraction of 2.5 Gy was given to the wrong patient. In other accidental exposures patients have been treated at the wrong anatomical site as a result of a variety of reasons, including using the wrong chart, using a tattoo port marker from a previous treatment, and reliance on asking the patient to identify the site. In the latter circumstance, the patient received a brachytherapy treatment with a strontium plaque to the eye instead of the prescribed 10 Gy external beam treatment; irradiation was initiated with the settings for rotational therapy from a previous patient; a technologist continued treatment with more fractions than prescribed.

(90) Contributory factors were:

(i) Procedures for identification of patients and the correct chart were not followed, including lack of verification of the treatment site against anatomical marks on the patient, and patient objections about being treated on the wrong site were not thoroughly investigated prior to treatment.
(ii) The oncologist relied only on asking the patient for the treatment site.

4.2 Brachytherapy

4.2.1. Equipment problems

(91) In the most serious case, with fatal consequences, the source became detached from the drive mechanism of a high dose rate (HDR) machine. In another equipment

failure, the wrong site was treated because a kink in the catheter prevented the source from reaching the correct position.

(92) Contributing factors in brachytherapy accidents due to equipment problems were:

(i) Equipment was not tested for reasonably foreseeable conditions.
(ii) Conflicting signals were misinterpreted (equipment indicated 'source shielded', but area monitor detected radiation) and the wrong indication was accepted.
(iii) Previous radiation monitor malfunctions encouraged misinterpretation and induced the staff not to trust the radiation monitoring device.
(iv) The patient, clothes, and room were not checked with a radiation monitor to establish the location of the radiation source.

4.2.2. Source ordering, delivery, calibration, and acceptance

(93) In one accidental exposure, the hospital and manufacturer used different units of activity (mCi and mg-Ra-equivalent respectively), and three cases of underdosage occurred because of failure to check the actual source activity. In another accidental exposure uncalibrated sources were used for many years, affecting many patients, with dose deviations of −5% to −29%.

(94) Contributory factors were:

(i) Delivery of 'wrong' sources by the supplier, i.e., the sources did not correspond to the accompanying source certificate.
(ii) Use of different units for source activity.
(iii) Lack of verification of the documents on source delivery against the documents of the purchase order.
(iv) Lack of verification of source strength before use.
(v) Interchangeable use of sources without consistency checks.
(vi) Usage of incorrect conversion factor on switching of sources from ^{226}Ra to ^{137}Cs.

4.2.3. Treatment planning

(95) Errors in this area have included incorrect time calculations, resulting in dose deviations ranging from −59% to +49%. Contributing factors were:

(i) Copies of an obsolete form were still available for clinical use.
(ii) Treatment plan did not include the reference point for the time calculation of a brachytherapy treatment.
(iii) Miscommunication between radiation oncologist, physicist, and dosimetrist (a treatment plan was modified but the unmodified plan was used).
(iv) Lack of independent time calculations.

4.2.4. Source preparation

(96) Incorrect or defective source use has caused some accidental exposures. In one case there was −50% of the prescribed dose, and doses were lower than intended over three months in another case. In other different cases, a manufacturer had delivered a source with essentially no activity; two ^{192}Ir sources were lost (they were separated from the ribbon and left unsecured), a leaking ^{125}I source was re-used, and sources that had been withdrawn from clinical use were used with an incompatible applicator. Factors contributing to these accidents were:

(i) Personnel handling sources and applicators lacked proper training.
(ii) Sources withdrawn from clinical use were not removed and were re-used by mistake.
(iii) Lack of verification of source activity.
(iv) Leakage of a source was not detected during preparation, and a similar incident that had occurred in another hospital had not triggered verification.
(v) Failure to identify the correct source ribbon end.
(vi) Lack of survey of all radiation sources before implantation.

4.2.5. Treatment delivery

(98) Examples of this type include: a junior physician who failed to implant one of the prescribed sources; the wrong patients were treated; a source ribbon was dislodged from the catheter and a nurse taped it on the face of the patient; sources were removed by a patient; and a source that did not match the applicator was loose and fell out of the applicator.

(99) Contributing factors were:

(i) An untrained physician worked without supervision; ordinary nurses with no special training were nursing brachytherapy patients; there were no written procedures; and poor communication of instructions that were not understood.
(ii) A prescription was misunderstood.
(iii) Lack of procedures for accountability of the sources.
(iv) The wrong chart was left on the console of the remote afterloading machine and used without verification.
(v) Errors were made with infrequently used treatment protocols.

4.2.6. Source removal

(100) In addition to the case described under 4.2.1 (HDR sources left inside the patient), there were several cases in which sources were lost due to lack of radiation monitoring after presumed source removal.

Contributing factors were:

(i) Sources were not accounted for after removal.

(ii) The patient, clothes, and room and/or waste from the treatment room were not monitored.
(iii) In one case, the sources were checked after removal against the total number of sources implanted, but not against the total number of sources that were sent to the room (more sources were sent than needed).

4.3. Accidents involving public exposure and environmental contamination

(102) The unsecured long-term storage of radiotherapy sources has led to catastrophic accidents with severe exposure and deaths among members of the public. These occurred as a result of hospital management, source suppliers, and importers relinquishing the accountability for sources.

(103) Contributing factors were:

(i) Non-compliance with regulations for transport and/or import.
(ii) Poor storage conditions before commissioning, or after decommissioning, of significant radiation sources.
(iii) Provisional short-term storage arrangements turning into long-term storage.

4.4. Generic lessons learned

(104) In most of the accidents, a combination of contributing factors allowed an initial mistake to escalate into an accidental exposure. In some cases this resulted in very serious or fatal consequences.

(105) Often, the lack of concern of management was the underlying root cause. When this situation is present it leads to many contributing factors including: lack of appropriate staff resources; insufficiently qualified or untrained staff; lack of effective, systematic quality assurance programmes/procedures, and lack of effective communication procedures. Often the need for reassessment of staff, resources, and training were ignored when a new machine was purchased or a new technology was introduced, or when the workload increased. Hospital management, source suppliers, and importers can cause catastrophic accidents involving the public and severely affecting the environment by relinquishing accountability for sources. These cases usually violate transport, import, storage, and decommissioning regulations.

5. RECOMMENDATIONS FOR THE PREVENTION OF ACCIDENTAL EXPOSURE IN RADIOTHERAPY

(106) It has been emphasised in multiple publications (cf. Hanks, 1984; WHO, 1988; AAPM, 1994; Thwaites et al., 1995; ESTRO, 1995; Alletti and Bey, 1996; IAEA, 1997; 1998a) that systematic quality assurance programmes in radiotherapy (QART) can prevent systematic errors and decrease the frequency and size of random errors. Prevention of most accidental exposures can be assured with minimal effort and expense in a radiotherapy department when two conditions are fulfilled: (i) a comprehensive and coherent quality assurance programme is in place, and (ii) for external beam radiotherapy, some in-vivo dose measurements are routinely performed.

(107) The first quality assurance programmes in radiotherapy dealt only with the verification of mechanical and electrical parameters of equipment and of dosimetric data. QART programmes have been extended progressively to include the verification of treatment planning procedures, patient set-up, and treatment delivery. Recently, the concept of quality assurance in radiotherapy has been extended to the verification of the entire radiotherapy process, from treatment prescription, patient data acquisition, and target volume delineation, to patient follow-up and treatment records. Some modern QART programmes also include recommendations for the structural organisation of a radiotherapy department and the qualifications and training of its staff.

(108) It is advisable that countries enforce regulations requiring radiotherapy departments to implement a comprehensive quality assurance programme. Some countries have adopted regulations on quality assurance in radiotherapy, making it compulsory to verify beam calibration by an external audit using mailed dosimeters (France) or have implemented mandatory periodic quality audit site visits, where independent calibrations and other type of measurements are made (Finland). However, it is important to emphasise that the development of QART programmes and the verification of their correct application is the responsibility of managerial and professional staff, mainly radiation oncologists and medical physicists. Radiotherapy technologists and dosimetrists have an important role in the application of the programme.

(109) It is not the purpose of this Chapter to provide detailed guidance on the contents of a quality assurance programme for radiotherapy, which are to be found in numerous publications[11] (c.f. WHO, 1988; AAPM, 1994; ESTRO, 1995; Alletti and Bey, 1996; IAEA, 1998a). Existing QART programmes provide comprehensive lists of parameters to be checked periodically, specifying the tolerance of the results measured and recommending the frequency of various checks, etc. They also emphasise the importance of double checks and independent verifications that provide defence in depth.

(110) The prevention of accidental exposures does not require an increase in the frequency or the type of checks, beyond the normal recommendations given in modern

[11] Most of the QART programmes are in accordance with guidelines provided by the ISO 9000 family of standards for quality assurance (cf. ISO), 2000) and with IEC standards for radiotherapy equipment (IEC, 1988; 1989; 1993; 1997; 1998; and in press), which have been extended to include recommendations on the proper use of equipment by the radiotherapy staff.

QART programmes, but rather the systematic application of the recommendations of a given programme. QART programmes are generally designed to detect minor errors and, as a consequence, they can also prevent the occurrence of major errors. The majority of the accidental exposures in radiotherapy have occurred in departments without a quality assurance programme or, when QART programmes existed, they were not fully implemented (omitting, for example, some of the checks to be performed).

(111) The aim of this Chapter is to provide recommendations of particular importance for the prevention of accidental exposures. The text follows the normal sequence of the radiotherapy process. Recommendations on staff functions, on equipment-related requirements and patient workload requirements for staff resources, and on the requirements for staff qualifications, can be found in numerous publications issued by national and international organisations (cf. WHO, 1988; IAEA, 1998a; ISCRO, 1991 for USA; ESTRO, 1996 for Europe). These can be adapted, if necessary, to the structure available in many countries. For the purpose of emphasising certain aspects of the staff responsibilities that will help to prevent accidental exposures, a summary of these responsibilities in relation to the education and training of the radiotherapy staff is included in this Chapter.

5.1. Structural organisation

(112) Radiation therapy is a multidisciplinary speciality, which uses complex equipment for the delivery of treatments to patients. A comprehensive quality assurance programme has clinical, physical, and administrative components whose implementation requires teamwork from all the professionals involved in the radiotherapy process.

(113) Key staff functions include radiation oncologist, medical physicist, radiotherapy technologist, and dosimetrist. Other categories of staff members that may be associated with a radiotherapy department, or may participate in the different steps of radiotherapy procedures, are mould room technician, nurse, maintenance engineer, etc. Special consideration should be given to staff workload, which may become excessive in the case of installation of new equipment, set-up of a new technology, or occasionally an increase in patient load. This will lead to an increased risk of accidental exposures, as a result of the difficulty of fully applying the quality assurance programme in these situations.

(114) The structure of a radiotherapy department should be clearly defined, especially with regard to the role and responsibilities of each staff member. In addition, the relationships between staff members and groups of professionals, either hierarchic or functional, should be clarified so that each individual understands her or his own position in the structure, as well as that of others. All decisions affecting the structure of the radiotherapy department should be recorded and the information should be disseminated effectively and quickly throughout the department.

5.2. Education and training

(115) A large number of accidental exposures have occurred because of the lack of qualified and well-trained staff. The most important component of the entire

radiotherapy process is qualified personnel. It is vital that all the staff dealing with radiation sources and patients have the necessary educational background and specialised training. Investment in equipment without concomitant investment in training is dangerous. Training should not only include practical details of individual procedures, but also the design of treatment approaches, which are comprehensive, reproducible, of high quality, and safe. Training should be consistent with the responsibilities assigned to each group of professionals, and should include the review and analysis of typical accidents together with a description of methods of prevention. There should also be continuing professional development programmes. Suggested guidelines for the education of staff members involved in radiotherapy are given below (cf. IAEA, 1998a) together with some aspects of their responsibilities related to accident prevention.

5.2.1. Radiation oncologists

(116) The physician practising radiation therapy should first be trained and experienced in oncological practice with post-graduate training in radiation oncology. The radiation oncologist will set the overall treatment policy for the radiation therapy programme and should participate in the design of the facility and the procurement of equipment. For individual patients, the radiation oncologist is responsible for the patient's care, including the details of the treatment and the patient's follow-up evaluation.

(117) The radiation oncologist who also practices brachytherapy should first be trained as a radiation oncologist and experienced in oncological practice. He/she should also have specific training in brachytherapy at an institution with an established practice, so that the indications for patient selection, applicator insertion, and dose prescription can be learned under the supervision of experienced mentors. The length of this training will depend on many factors, but will usually be measured in months. Such a training period should be undertaken whenever a substantially new form of brachytherapy is introduced into an existing practice, for example when adding high dose rate brachytherapy.

5.2.2. Medical physicists

(118) The medical physicist should have at least an advanced university degree in a physical science or engineering, at least one year of academic and clinical training in radiation oncology physics, and additional training of at least one month in brachytherapy physics at an established centre, preferably the same centre as that visited by the radiation oncologist, if these treatments are to be undertaken. In this manner, a consistent and comprehensive practice can be developed. It is also very helpful for a junior physicist to have at least a part-time training in a department with senior physicists. Special training courses are currently available for medical physicists and are provided by international bodies in different parts of the world.

(119) The medical physicist's responsibilities cover four major areas: dosimetry, equipment acceptance and commissioning, quality control, and radiation safety. In

dosimetry, the medical physicist helps minimise the probability of patient injury and poor treatment outcome by assisting in devising, for each patient, an appropriate treatment regimen, and reviewing all patients' treatment plans. The medical physicist is responsible for the calibration of the output of the treatment machine, during commissioning and on a routine basis, and for assuring that all machine data used for patient treatment is accurate and adequate. For quality control, the medical physicist will be involved with establishing and running a quality control programme which includes patient safety. Radiation safety also requires the establishment and maintenance of a radiation protection programme designed to ensure the safety of staff and the public. These duties will be the responsibility of the medical physicist, and/or the Radiation Protection Officer who may or may not be the same person. The administrative structure will vary depending on the nation, the facility, and the resources; what matters is that the necessary authority is available.

(120) It should be understood that for the practice of radiation therapy is it mandatory that a hospital have access to a suitable medical physicist. It is not sufficient that the physics staff be trained; they should also be available in sufficient numbers to carry out all the required duties.

5.2.3. Radiation therapy technologists, dosimetrists, and nurses

(121) In addition to the radiation oncologist and the medical physicist, a radiation therapy programme requires radiation therapy technologists, dosimetrists, and radiation oncology nurses. Radiotherapy technologists have the responsibility for the set-up and delivery of the treatment, are involved in the simulation of the treatment, and have, therefore, an essential function in noticing any abnormal reaction of the patient or the machine and to report them as indicated in Section 5.4. Thus, radiation therapy technologists play an important role in preventing accidents.

(122) Radiation therapy technologists, dosimetrists, and nurses should have a degree, granted by a university or medical school, in academic studies and clinical training for a period of three or four years. Nurses in charge of patients during brachytherapy treatment preparation and/or treatment delivery should receive training that emphasises the importance of technical features (e.g., the correct position of sources). In addition, they should have clear instructions on the necessity to report immediately unexpected events to a medical physicist, a radiation oncologist, or a radiation protection officer.

(123) Although the radiation oncologist and medical physicist may delegate specific duties to these personnel as appropriate, they will retain the responsibility for providing adequate supervision and training. For example, computerised dose calculations may be performed by a 'treatment planner dosimetrist', or preparation of low dose rate sources for patient treatments and maintenance of the source inventory may be delegated to a 'source curator'. Such individuals can perform valuable service as technicians, especially where more highly trained persons are scarce, but they should not be given responsibilities beyond their professional competence.

(124) A clear delineation of responsibilities is particularly important in the case of dosimetrists and technologists. In some institutions, these professionals substitute

for medical physicists, and treatment planning and delivery procedures are made without the supervision of a qualified medical physicist. Whether this is done for economic or practical reasons, such practices might have detrimental consequences for the patient. For example, the lack of education in the specialised areas of mathematics and physics restricts the understanding of the algorithms used in modern computerised treatment planning systems; this can easily jeopardise the interpretation of results produced by limitations of a treatment planning system. Furthermore, although the competence of a dosimetrist may be adequate for most problems found in typical routine work, their training may not be sufficient to identify the causes of abnormal or unexpected situations, and to decide which checks are required and their degree of urgency.

5.2.4. Maintenance engineers

(125) If there is a large amount of equipment, e.g. several external therapy units and simulators, block cutting equipment, treatment planning computers, tissue compensation devices, etc., it might be advisable to ensure the immediate availability of trained engineering maintenance staff. If remote after-loading devices are used in the brachytherapy programme, then provision should be made for servicing the devices. This may be best accomplished with service agreements with the manufacturer. Alternatively, staff will need to be trained in repair and preventative maintenance of equipment, as well as in the basics of radiation protection.

5.3. Acceptance testing and commissioning of equipment

(126) A number of important steps should be taken before, during, and immediately after radiotherapy equipment is delivered to a facility. Radiation sources need to be safely received, registered, and stored, the radiation measurement equipment tested and calibrated, the shielding of special rooms measured, the radiation sources tested and calibrated, and teletherapy units and remote afterloaders commissioned. A record-keeping system should be in place.

(127) The acceptance tests demonstrate whether the equipment meets or exceeds the procurement specifications. Frequently, acceptance tests follow a protocol supplied by the manufacturer, but the purchaser may develop his/her own protocol. In either case, the acceptance test protocol should be part of the purchase order for the equipment. Acceptance tests protocols specify what tests will be performed, what equipment is used to perform these tests, and what the results of these tests should be. They constitute a legal document in which the medical physicist confirms that the equipment met the procurement specifications.

(128) Acceptance tests are crucial to the prevention of accidental exposures, because safety interlocks are tested for the first time in the hospital, and because some of these interlocks cannot be tested under normal operation. At this stage, once the installation has been completed, for the first time it is possible to test the safety of the system as a whole, including interaction between different parts of the equipment. Acceptance tests have been facilitated by the introduction of international standards for medical electrical equipment (such as the IEC or equivalent national standards), which have

started to specify the tests to be performed at the hospital level and the methods of testing for the user. For those interlocks that cannot be tested at hospital level, these standards also specify the type of evidence that the manufacturer has to provide of the testing in factory. It is therefore essential that the hospital staff responsible for the acceptance, usually the medical physicist, be acquainted with these standards.

(129) At the completion of acceptance tests, commissioning measurements begin. During commissioning measurements, the physicist will measure all the data required to place the unit into clinical service. The physicist should ensure that all data needed to perform any anticipated clinical procedure is acquired at this time. The data should be acquired in the format required for entry into the treatment-planning computer. All data should also be entered into a logbook for archival purposes. The pages of the logbook should be dated and signed by the physicist. Formal written acceptance should only be signed after these tests have been performed.

(130) Immediately after the conclusion of the commissioning measurements, quality control tests should be established, following the quality assurance programme adopted by the institution. Checks should only be performed by qualified and experienced persons, such as the medical physicist, who can sometimes delegate the work to persons she/he has trained. Regardless of who performs the tests, the medical physicist remains the responsible party for assuring the correct performance of the equipment. The medical physicist should also verify that the data in the treatment planning computer, any computer used to calculate treatment times, and in the logbook, are correct and consistent.

5.4. Follow-up of equipment faults

(131) The safe repair of equipment is essential in the prevention of accidental exposures. As shown in the case histories, some unsafe repairs, or lack of follow-up of unexplained malfunctions, ended in fatal accidents. The most difficult situations are those malfunctions that cannot be reproduced by the maintenance engineers, either due to intermittent faults or faults that only appear in some particular circumstances. In these cases, it will be difficult for the maintenance engineer to isolate the cause and to remove it. A procedure that facilitates interaction between users and manufacturer, a follow-up of unexplained malfunctions, and dissemination of the information to other users of the same type of equipment and to maintenance engineers is the recommended approach to maintenance.

5.5. Communication

(132) Many accidental exposures could have been avoided by better communication between the different staff members or between staff and maintenance engineers. Communication should be assured in the organisation of the department and, for instance, a maintenance engineer should not be allowed to work on a treatment unit without clearance by those responsible for a unit. Most of the problems of communication identified in some of the accidental exposures reported in this document could have been solved by using one of the following procedures:

(i) If anomalous behaviour is observed in a treatment unit, this should be reported immediately to the medical physicist. If the behaviour is suspected to be caused by a change in machine performance, the physicist should carry out suitable checks for assessment promptly.
(ii) If an unexpected reaction is observed in a patient by a radiotherapy technologist, or indicated by the patient, the radiation oncologist should immediately request the medical physicist to perform a verification to detect any possible error in the different steps of the treatment procedure.
(iii) If unexpected reactions are observed for multiple patients, the medical physicist should be informed immediately and requested to verify the dosimetry of the treatment unit.

(133) In any of these circumstances, the medical physicist should recommend the radiation oncologist not to continue patient treatments until equipment performance or treatment parameters have been verified. A heavy workload on a treatment unit should never be a reason to resume patient treatments without the necessary verifications.

(134) It is essential that every radiation treatment be carefully documented. Treatment charts, radiographs, and all patient documents should be clearly identified and dated to avoid confusion between patients or between different stages of the treatment for one patient. If an accident occurs, these documents will be essential for the radiation oncologist in any decision relative to further patient care.

5.6. Patient identification and patient chart

(135) Some of the accidental exposures resulted in the delivery of a treatment or part of it to the wrong patient, or in treating the wrong site. Errors with the patient chart resulted in mistakes in the number of fractions or the dose or with the changes in the course of the treatment, for example, new fields, or prolonging the treatment beyond the plan. It is important that the quality assurance programme includes provisions and procedures for an effective identification of the patient (photograph, ID) and site, and that a chart check protocol is implemented. The chart review should be at least once a week, before the third fraction following the start of a new treatment field or modification and at completion of treatment (AAPM, 1994).

5.7. External beam radiotherapy

(136) In external beam therapy most accidental exposures involving a large number of patients have occurred in relation to erroneous calibration of the beam and, to a lesser extent, with the calculation of the dose (or treatment time) to be delivered. Other accidents involving one patient or a few patients were related to simulation and treatment delivery.

5.7.1. Beam calibration

(137) The calibration of a radiotherapy beam using a well established dosimetry protocol is mandatory, whenever a new machine is installed or the source of a ^{60}Co

teletherapy unit exchanged. Dosimetry protocols like IAEA TRS-277 (IAEA, 1987) and AAPM TG-21 (AAPM, 1983) can be recommended for air kerma-based dosimeter calibrations whereas IAEA TRS-398 (IAEA, 2000b) and AAPM TG-51 (AAPM, 1999) can be used for absorbed-dose-based dosimeter calibrations. Attention must be paid to the correct understanding of the type of dosimeter calibration (N_X, N_K, $N_{D,w}$) prior to the selection of a dosimetry protocol. Periodical verification measurements should be made in accordance with the recommendations of the quality assurance programme adopted.

(138) A redundant independent measurement, to verify the calibration, should be performed before the first patient is treated. It is recommended that national regulations require independent calibrations for all radiotherapy treatment units and that regulations be enforced. The redundant calibration can be done by staff of another radiotherapy department, by an external audit using a postal dosimetry service (for example with TLD, see Appendix B) or by a site visit conducted by an independent organisation.

(139) The results of the redundant calibration should not differ from the original calibration by more than 2 to 3%, depending on the type of detector used. If the difference is more than 5%, the reasons for the discrepancy should be carefully investigated and patient treatment not initiated until the discrepancy is resolved. With discrepancies around 3–5%, patient treatments can start, but still the reasons for the deviation must be identified. It is essential to understand that the purpose of independent verifications made with TLD mailed dosimeters is to discover possible errors in the calibration, but these should never be used as a substitute for a proper local calibration, using an instrument with a valid traceable calibration certificate and using a widely accepted dosimetry protocol.

(140) Every radiotherapy centre should participate regularly in an external audit programme to verify the calibration of treatment units, ideally with the periodicity of one year, but not less frequently than every five years. It has been reported (Dutreix et al., 1993) that the size and number of discrepancies in beam calibration in centres that have participated regularly in external audits is much smaller than for centres that have not participated in such programmes.

(141) Quality control checks, particularly the verification of the dose delivered in reference conditions, should be made at regular intervals. In addition, the calibration of a treatment unit should be performed after any repair or adjustment of the unit that may affect the characteristics of a beam. Other parameters related to dose delivery in clinical conditions should be checked regularly, following in detail the quality assurance protocol adopted by the institution.

5.7.2. Treatment planning systems

(142) Severe accidents in radiotherapy have been associated with the incorrect commissioning of computerised treatment planning systems. Careful verifications of absolute and relative doses delivered should be performed, including in-phantom measurements for a comprehensive number of typical configurations. These are included in the majority of quality assurance protocols for treatment planning sys-

tems (cf. ICRU, 1987; AAPM, 1998). The physical basis on which the algorithms in the treatment planning system are based should be clearly understood; this recommendation applies also to any additional correction factors introduced manually in dose calculation procedures.

5.7.3. In-vivo dose measurements

(143) Many of the accidents described in this publication could have been avoided if in-vivo measurements had been performed on a selected group of patients. In-vivo measurements (Leunens et al., 1990; Garavaglia et al., 1993; Van Dam and Marinello, 1994) are an effective way of verifying the quality of the entire radiotherapy treatment procedure. The additional cost of in-vivo dosimetry does not require a considerable increase in funding even in a small hospital (Kesteloot et al., 1993). It is an especially valuable investment, but to be effective, it requires careful preparation in terms of equipment, staff training and quality assurance.

(144) Diodes and thermoluminescent dosimeters can be used for in-vivo measurements. It is important to realise that when the detector used for in-vivo dosimetry has been calibrated in the same treatment unit where patients are treated, the results from in-vivo measurements will be correlated with the calibration of the machine and, therefore, will not be able to show a potential error in the calibration of the machine. A correct calibration of the dosimeter is thus an essential necessity.

(145) Patient entrance dose measured on the central axis of the incident beam, performed systematically on each patient during the first session of each field, provides information on a combination of parameters from treatment preparation to treatment delivery. Combined with exit dose measurements, they provide information on patient data and on the performance of the treatment planning system used. When results are carefully analysed large systematic errors can be detected rapidly, even with a limited number of measurements.

5.8. Brachytherapy

(146) The majority of the accidental exposures that occurred with brachytherapy treatments can be linked to source parameters, to dose calculation procedures and to insufficient training of personnel.

5.8.1. Source activity and identification

(147) The activity of each brachytherapy source should be checked individually, before it is used on a patient. Some of the major accidents in brachytherapy have been caused by errors in the manufacturer's specification of the activity of one or several sources. Radiation oncologists should never use a source until its activity has been verified. The unit of activity used at the hospital may differ from the unit stated by the manufacturer. The documentation should be checked carefully. It is essential that the unit of activity used for source calibration be the same as the unit of activity used in the treatment planning system. The international recommendations for the

specification of source strength (cf. ICRU, 1985) should be implemented for all brachytherapy sources. After verification of the source activity, the source or source holder should be marked with unique identifiers, to prevent the possibility of confusion between different sources.

5.8.2. Dose calculation and treatment planning

(148) The calculation of the decay of the source is not always incorporated into the dose calculation software, and a calculation by hand to take into account this effect may be necessary for each patient. As information on the nature of the sources for which the software has been developed is not always clearly stated, it is essential that this radionuclide be identified. The use of radionuclide specific software with a different radionuclide may lead to severe errors on dose calculation and to accidental exposures. Dose specification should be done following the recommendations issued by international bodies (e.g., ICRU, 1985; 1993; 1999).

5.8.3. Source positioning and removal from the patient

(149) It is essential to verify that a source has been positioned correctly in the patient and remains so during the treatment time. In remote afterloading devices where sources are inserted automatically several times, the reproducibility of the position should be checked before each application. Patients should be informed of the necessity to avoid movements that may cause a source displacement during the treatment. In some hospitals, brachytherapy treatments are not performed on patients who are unable to understand these requirements, unless there is a way to control patient movements. When a brachytherapy treatment is completed, the sources should be removed from the patient either automatically or manually, depending on whether a remote afterloader device is used. The patient and the patient's bedroom should be checked carefully with a simple survey monitor to assure that all sources have been removed and properly stored.

(150) Fig. 5 summarises the recommendations given in this Section.

5.9. The potential for accidental exposures in the future

(151) The recommendations included in this Chapter are based on a retrospective analysis of accidental exposures in radiation therapy with past and current types of equipment. There are, however, a number of factors that may cause a change in this picture in the future:

- With the worldwide expansion of radiotherapy there may be more accidents related to inadequate staff training, mainly in countries where education programmes are still not widely implemented.
- There is a common misperception that modern equipment is safer and will require less quality assurance.

Checklist for the Prevention of Accidental Exposures

Organisation, functions, and responsibilities
- Have all necessary functions and responsibility been allocated?
- Are all functions and responsibilities understood?
- Is the number of staff commensurate to workload?
- Is this number re-assessed when workload increases, or when new equipment is purchased?

Education and training
- Is every member of the staff educated and trained according to their responsibilities?
- Is this education and training documented?
- Is there a programme for continuing and personal development?
- Are lessons from accidents and their prevention included in continued training ?
- Are there provisions for additional training (new equipment, new procedures)?
- Are emergency plans exercised as part of the training?

Acceptance testing and commissioning
- Is there a programme for formal acceptance of equipment in place?
- Is it carried out according to international or national standards?
- Is there a programme of commissioning in place?
- Does it include treatment equipment as well as treatment planing systems and simulators and other ancillary equipment?

Quality Assurance Programme
- Is a programme of QA established?
- Is the programme based on accepted protocols? Which ones?
- Are all tasks of the QA clearly assigned to the right persons?
- Are the necessary tools and instruments available?
- Are audits part of the programme?

Communication
- Is a communication policy in place and understood by the staff?
- Is reporting of unusual equipment behaviour required?
- Is reporting of unusual patient reactions required?
- Are procedures for equipment transfer for maintenance and return in place?

Patient and site identification
- Are there procedures to ensure correct identification of patient and site?
- Is there a protocol for patient's chart check?

External beam

Calibration
- Are there provisions for initial beam calibration?
- Is independent verification in place foreseen and planned?
- Is there an accepted protocol? Which one?
- Is a programme for follow-up calibration in place?
- Is participation in an audit programme part of the programme?

Treatment planning (clinical dosimetry)
- Are treatment planning systems included in the programme of acceptance and testing?
- Is treatment planning documented according to accepted protocols?
- Are cross-checks and redundant and independent verification included?

In-vivo dosimetry
- Has a system for in-vivo dosimetry been considered?

Brachytherapy

Source activity and identification
- Are there provisions for source activity verification and identification of the source before use?

Dose calculation and treatment planning
- Are there provisions for dose calculation and cross-checks?

Source positioning and source removal
- Are there provisions to verify source position and to ensure that position remain?
- Are there provisions to ensure that sources do not remain in the patient, including monitoring of patients and clothes?

Fig. 5. A checklist for accident prevention.

- An increasing number of clinical accelerators are being installed, either as new therapy units or as a replacement of ^{60}Co teletherapy machines, many of them in developing countries. Accidents may occur due to inadequate accelerator maintenance, as many developing countries may not have the infrastructure required for the maintenance of complex equipment. The increased number of computer-controlled systems may also lead to more computer related accidents, compared to mechanical failures.
- The increased use of accelerators will require the decommissioning of ^{60}Co teletherapy units and perhaps some old ^{137}Cs units, which may result in an increased accident rate due to improper disposal. After a decade or so the frequency of these events will probably decrease.
- The new technologies of high dose rate (HDR) brachytherapy, 'gamma knife' therapy units, multi-leaf collimators, and intensity modulated conformal radiotherapy (IMRT) may produce new types of accidental exposures[12].

(152) The consequences of these factors and new types of events can only be anticipated by fault tree analysis. However, such analysis may not be able to take fully into account human factors (e.g. education and training). It is expected that an improved education/training system and enhanced quality assurance culture, together with carefully designed regulations and adequate enforcement, will be the seed to prevent the occurrence of severe accidents in the field of radiation therapy.

[12] Note that in some countries neurosurgeons with little or no training on radiation effects, physics, or protection are the major operations of 'gamma knife' units. A similar situation occurs with endovascular brachytherapy sources, which often are manipulated by cardiologists.

APPENDIX A. UNCERTAINTY IN RADIOTHERAPY

(A1) The physical quantity 'absorbed dose', like any other quantity, coefficient or parameter, has an uncertainty associated to its numerical value that characterises the dispersion or spread of values determined around an estimated value of the quantity. This is described by a parameter like the standard deviation, which reflects the lack of exact knowledge of the value of the quantity. Note that a result can have a negligible error (be very close to a reference value), but yet it may have a large uncertainty (ISO, 1995). The overall uncertainty in a process is obtained from the combination of the uncertainties in all the steps involved in the process, yielding the so-called combined standard uncertainty, u_c (ISO, 1995).[12]

A.1. The influence of dose uncertainty in radiotherapy treatments

(A2) Uncertainties in absorbed dose arise during all steps of a treatment, from beam calibration to dose delivery. The dispersion or spread of dose values received by patients (Fig. A.1) is assumed to be a Gaussian distribution around the prescribed dose. The width of the distribution is given by its standard deviation or standard uncertainty. When a mistake or a systematic error causes a difference

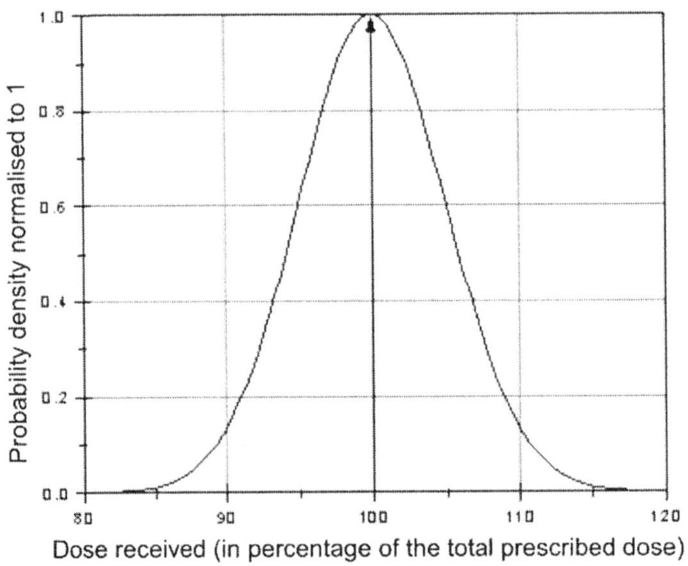

Fig. A.1. Gaussian function representing the dispersion of the dose received around the prescribed dose value (100%).

[12] An expanded uncertainty can be obtained using a coverage factor k, $U = k\ uc(k = 1, 2, 3)$. With $k = 2$, i.e. two standard deviations, the expanded uncertainty corresponds approximately to the conventional 95 % confidence limit.

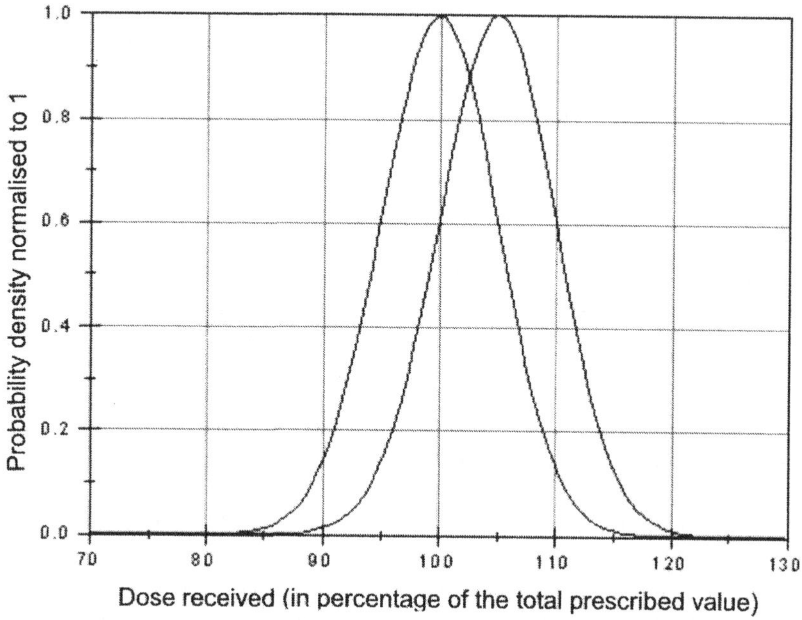

Fig. A.2. Prescribed and delivered doses differing by 5%, each having an associated uncertainty described by a Gaussian distribution of width 5%. The width of the distributions poses a limit to the 'resolution' with which the two distributions can be observed as clearly separated.

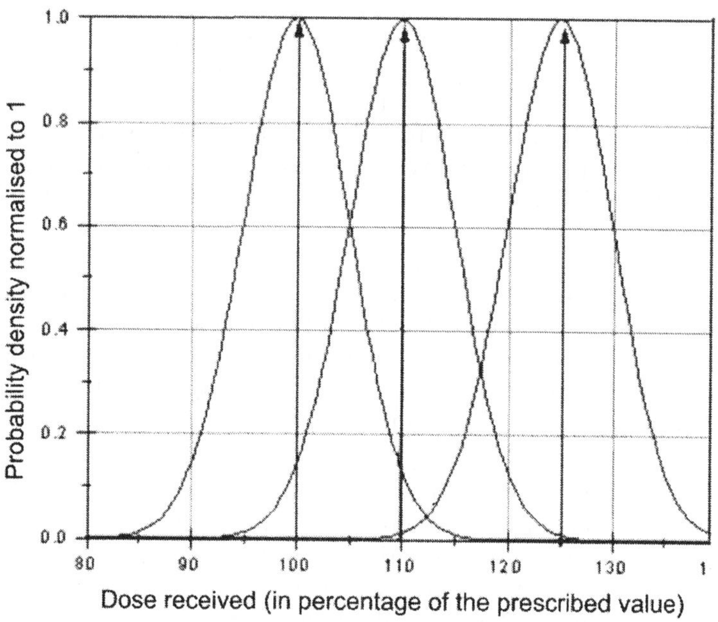

Fig. A.3. Gaussian distributions illustrating dose differences of 10% and 25%.

between the dose received and the dose prescribed, the distribution of doses to the affected patients is shifted away from the prescribed value (Fig. A.2).

(A3) As shown in this Appendix, the combined standard uncertainty in good radiotherapy practice is of the order of 5%, i.e., approximately of the same order as the lowest dose differences that can be detected clinically. The smaller the differences between the dose received by patients and the dose prescribed, the more difficult it is to detect errors in clinical practice, and they can only be detected when a relatively large number of patients are affected and carefully observed. Dose uncertainties thus pose a limitation to the 'resolution' in detecting an error in dose. When the differences between the dose received by patients and the dose prescribed are large, problems in the clinical outcome become more noticeable, as the distributions are more separated (Fig. A.3).

A.2. UNCERTAINTY IN EXTERNAL BEAM RADIOTHERAPY

(A4) Following the work by Loevinger and Loftus (1977), estimates of the combined standard uncertainty of the dose distribution in the target volume, along the different steps of a treatment with external beams, have been given by Dutreix (1984), Svensson (1984), and Brahme et al. (1988). These have been updated, mainly with regard to the calibration process, by the evaluations of Andreo (1990, 1992) and Thwaites (1994).

(A5) Uncertainty estimates are summarised here where the various steps of a radiotherapy treatment are classified into four major groups according to:

(i) absolute dose determination at the reference point in a water phantom (beam calibration), performed with a calibrated ionisation chamber according to a code of practice or dosimetry protocol;
(ii) procedures involving relative beam dosimetry (field size output dependence, influence of beam modifiers like wedges, secondary collimation, and blocks, etc.), performed with any type of detector suitable for measurements in water or in a plastic phantom;
(iii) the calculation of the dose delivered to the patient (monitor units or irradiation time to deliver the prescribed dose) and its distribution, usually performed with a computerised Treatment Planning System; and
(iv) the process of treatment delivery along a complete treatment, which accounts for daily variations in patient and machine set-up, patient movements, and machine instability during several weeks of treatment.

(A6) The uncertainties estimated for the various groups are given in Table A.1 for radiotherapy treatments in which high-energy photon beams produced by clinical accelerators are used, along with the various contributions to the uncertainty of each group. Under the heading 'Overall process' in Table A.1, the column lists all possible contributions to the combined standard uncertainty in each step. The resulting combined standard uncertainty is about 6 %.

Table 1. Estimated relative standard uncertainties (u_c) in the various steps of a radiotherapy treatment with high-energy photon beams produced by clinical accelerators. Estimates correspond to clinical environments where high performance in dose determination and treatment delivery can be achieved. Data marked as '-' in the rightmost column are considered to be practically identical within an institution or when several institutions adopt a common protocol for clinical comparisons; they do not contribute to observed differences in clinical results.

	Step in the radiotherapy treatment	Overall process		Clinical comparisions	
1.0	**Absolute dose determination at the reference point in a water phantom (beam calibration)**				
1.1	dosimetry recommendations (protocols, codes of practice, standards of practice, etc.)	2.0		–	
1.2	experimental procedure (including monitor errors)	1.0		1.0	
	combined standard uncertainty in step 1		*2.2*		*1.0*
2.0	**Relative dosimetry (water or plastic phantom measurements)**				
2.1	beam size output factors (including monitor errors)	1.0		1.0	
2.2	beam modifiers (wedges, blocks, etc.)	1.0		1.0	
2.1	measurement of isodose distributions	1.5		1.5	
	combined standard uncertainty in step 2		*2.1*		*2.1*
3.0	**Dose calculation to the patient**				
3.1	patient input data (contour, CT-data, etc.)	1.0		1.0	
3.2	Treatment Planning System (beam data, algorithms, inhomogeneity corrections, etc.)	3.0		3.0	
3.3	water to tissue transfer (including physical data)	2.0		–	
	combined standard uncertainty in step 3		*3.7*		*3.2*
4.0	**Dose delivery (along complete treatment)**				
4.1	reproducibility in beam set-up, beam alignment, etc	1.0		1.0	
4.2	reproducibility in patient set-up	2.0		2.0	
4.3	contour and organ motion	1.5		1.5	
4.4	machine instability	2.0		2.0	
	combined standard uncertainty in step 4		*3.4*		*3.4*
	combined standard uncertainty in all steps	**5.9**		**5.1**	

(A7) It is important to realise that not all the items appearing under the column 'Overall process' are included in published analyses of clinical results and estimates of accuracy requirements in treatment delivery. Several components correspond to steps whose uncertainties are dominated by those of fundamental physical quantities and constants; this is mainly the case with correction factors and coefficients given in dosimetry recommendations (protocols and codes of practice) for beam calibration[13], as well as for data to convert dose from water to various types of tissue.

(A8) Uncertainties in such factors will, in general, be practically the same in most clinical intercomparisons and they are assumed not to influence the results of relative

[13] The uncertainty assigned to the calibration of therapy machines is the main factor responsible for the differences in uncertainty estimates in radiotherapy treatment delivery published by different investigators (Svensson, 1984; IAEA, 1987; Mijnheer et al., 1987; Brahme et al., 1988; Andreo, 1990, 1992; Thwaites, 1994; etc).

clinical studies. Under the column 'Clinical comparisons' these components have been omitted, yielding a combined standard uncertainty of about 5%. It is possible to arrive at situations where still some other contributions may cancel out in a comparison, for example for treatment outcomes within a single institution, sometimes even with the same treatment unit. In those very special cases it is possible to arrive at a combined relative standard uncertainty of about 3.5%. On the other extreme, when beams other than high-energy photons are used (for example electrons), measurements with instruments other than reliable ionisation chambers are made, or patients are treated with sophisticated procedures, uncertainties (and the likelihood for errors in treatment delivery) are expected to increase, so that an estimate of 5% may be difficult to achieve. These cases are unfortunately not infrequent in routine clinical practice.

(A9) The conclusion from this analysis is that an estimated standard uncertainty of 5-6% seems to be a figure achievable at institutions with clinical practice of high standard for carefully delivered radiotherapy treatments, but it is difficult to decrease the combined uncertainty beyond this level. Studies assuming a narrower uncertainty, based on the assumption that contributions cancel out, can only be justified under very special circumstances.

Annals of the ICRP

APPENDIX B. QUALITY AUDITS OF THE CALIBRATION OF RADIOTHERAPY BEAMS

(B1) It is of interest to analyse the actual accuracy in dose determination in radiotherapy beams worldwide. The information presented in this Appendix is based on dose quality audits of radiotherapy centres (sometimes referred to as 'intercomparisons') performed with mailed thermoluminescence dosimeters (TLD) under the IAEA/WHO TLD postal service (Izewska and Andreo, 2000)[14]. Since 1969, this postal service has been used to check more than 3000 radiotherapy beams, and in many instances significant errors have been detected in their calibrations. These calibration errors have often resulted in patient mistreatment and some could even be included in the category of accidents. Fortunately, the number of substantial calibration errors has decreased during recent years.

(B2) Results for the period of 1990–1997 (about one third of the total number of beams checked during 30 years) are summarised in Fig. B.1, which reflects worldwide data, mainly from developing countries. There are no substantial differences between the various geographical regions. It can be seen that approximately 75% of the results correspond to beam calibrations with deviations within ±5% (previously this figure was 65%), leading to treatment outcomes most likely within clinical tolerance. Fourteen percent of the results have deviations between 5–10%. The remaining eleven percent of the results with deviations larger than 10% will have severe detrimental consequences, and of these, four percent of the beam calibrations can be considered associated with the likelihood of a radiotherapy accident.

(B3) Many severe discrepancies are solved through follow-up verifications and site visits by national or IAEA experts, although in some instances the lack of further reply by the institution has made it impossible to solve the problem detected. Of special concern is the large amount of dosimeters not returned for evaluation (approximately 25–30% of the dosimeters distributed, although the trend is decreasing; previously this figure was 40%). In many cases this corresponds to institutions not willing to show their level of performance. It is not known how many radiotherapy machines exist in the world that have not been involved in such quality audits.

[14] The IAEA operates a cost-free service to verify the calibration of radiotherapy units in hospitals and oncology centres. The TLD postal service is implemented through a collaboration between the IAEA and the WHO (PAHO in Latin America) and is known as the IAEA/WHO TLD postal service. In all instances the service provides an independent and impartial quality audit of the dosimetry procedures used at the hospitals. The TLD procedures receive the support of the Bureau International des Piods et Mesures (BIPM), various Primary Standard Dosimetry Laboratories, and some advanced radiotherapy centres and institutions in Europe and USA. The service warrants the confidentiality of the results, and only the persons responsible for the radiotherapy departments or for the calibrations have access to the outcome of the verification.

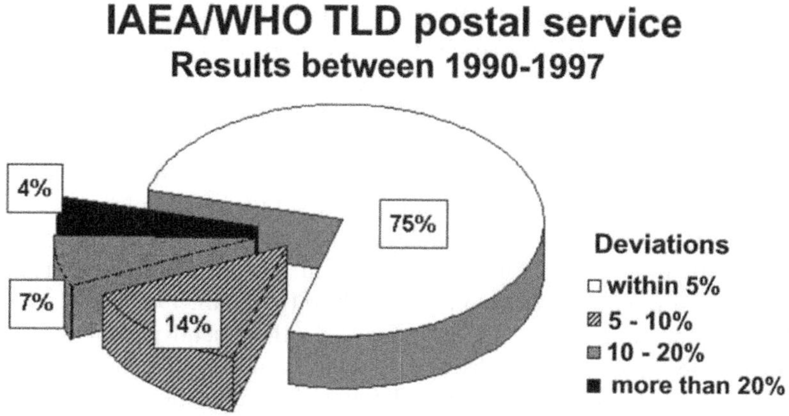

Fig. B.1. Results of the IAEA/WHO TLD postal service for the calibration of therapy beams during the period 1990–1997 (approximately 1100 beams verified). Measured deviations relative to the IAEA reference dose are grouped according to the levels in the legend. Data shown reflect worldwide figures, mainly from developing countries; there are not substantial differences between the various geographical regions. Approximately 25–30% of the dosimeters distributed by the TLD postal service were not returned for evaluation.

(B4) There are several other national and international organisations providing this type of inexpensive quality audit service to various countries and regions (cf. Hanson et al., 1997; Ferreira et al, 2000). This is probably the simplest method aimed at preventing radiotherapy accidents. These TLD-mailed procedures have been recently adopted by a number of international and national organisations (cf. IAEA, 1997). Participation in such quality audits is recommended in numerous protocols for quality assurance in radiotherapy as a complement to or replacement of the more efficient, but more expensive, site visits performed by an external audit group (cf. Hanson et al., 1997; Järvinen, 1997).

APPENDIX C. CASE HISTORIES OF ACCIDENTS RELATED TO DECOMMISSIONING OF RADIOTHERAPY EQUIPMENT AND SOURCES

C.1. Illegal import, storage, and disposal of a teletherapy ^{60}Co unit (Mexico, 1984)

(C1) A second-hand teletherapy unit with a 1,000 Ci (37 TBq) ^{60}Co source was purchased and imported, but did not comply with all the existing import regulatory requirements. It was stored in a warehouse for six years; then its scrap value attracted the attention of a maintenance technician, who dismantled the head of the unit and removed the cylinder containing the ^{60}Co (CNSNS, 1985).

(C2) The technician removed the cylinder and other metal parts from the unit, loaded them into a pickup lorry, drove to a scrap yard and sold the parts as scrap. Before arriving at the scrap yard, he deliberately ruptured the cylinder containing the source. The pickup lorry thus contained a large quantity of radioactive material from the source. The source consisted of about 6,000 tiny pellets (1 mm diameter) of ^{60}Co. When the cylinder was ruptured, several pellets were dispersed and remained in the lorry when the heavy parts were unloaded at the scrap yard. Owing to mechanical defects, the pickup lorry contaminated with ^{60}Co pellets remained parked on the street for 40 days. The lorry was then moved to another street where it stood for a further ten days.

(C3) When the ruptured cylinder was moved by cranes in the scrap yard, together with the other metal pieces, the ^{60}Co pellets were spread over the scrap yard, attracted by the magnetic field of the crane, and mixed with the other metal materials. Consequently, pellets and pellet fragments also were transferred to other vehicles used for transporting scrap to various foundries. The main purchaser of the scrap was a firm that manufactured construction reinforcing rods and connecting rods for motor vehicles.

(C4) It was later found that scrap contaminated with ^{60}Co had been used by steel production plants to manufacture reinforcing rods and metal table bases. A lorry transporting contaminated rods to be used for construction entered a nuclear laboratory where radiation detectors were used to monitor removal of radioactive material. The detectors not only indicated the presence of radioactivity, but also activated a camera that photographed the contaminated vehicle. Two days later the authorities ascertained the origin of the contaminated rods.

(C5) An extensive investigation showed that 30,000 table bases and 6,600 tons of reinforcing rods had been made from the contaminated material. Aerial radiation surveys of an area of 470 square kilometres were conducted, resulting in the recovery of 27 cobalt pellets. Visits were made to 17,636 buildings to determine whether contaminated material was used in their construction. Unacceptable radiation levels were measured in 814 buildings that were then partly or completely demolished. The accident exposed approximately 4,000 people to radiation, about 80 of whom received doses greater than 250 mSv. Apparently, five people received doses of 3 to 7 Sv over a 2-month period.

(C6) Although the triggering event was that a person dismantled an insecurely stored head of a ^{60}Co teletherapy unit and broke the source capsule, other factors contributed, such as non-compliance with regulations, as the radiotherapy unit was illegally imported and transported and remained in storage for six years under unsafe and unsecured conditions.

C.2. Abandonment of a teletherapy ^{137}Cs unit (Brazil, 1988)

(C7) A private radiotherapy institute moved to new premises, leaving a caesium (^{137}Cs) teletherapy unit without notifying the licensing authority. Because of partial demolition of the building, the teletherapy unit became totally unsecured. Two people entered the building and removed the source assembly from the radiation head. They tried to dismantle the source assembly at home and in the attempt, the source capsule was ruptured. Because the radioactive source was in the form of caesium chloride powder, which is highly soluble and easily dispensed, there was considerable contamination of the environment, resulting in external irradiation and internal contamination of several persons (IAEA, 1988).

(C8) After the source capsule was ruptured, remnants of the source assembly were sold for scrap to a scrap-yard owner. He noticed that the source material glowed blue in the dark. Several persons were fascinated by this, and over a period of days, friends and relatives came and saw the phenomenon. Fragments of the source the size of rice grains were distributed to several families. Five days later, a number of persons showed gastrointestinal symptoms that were not recognised initially as being due to exposure to radiation. However, one of the persons irradiated connected the illnesses with the source capsule and took the remnants to the public health department. This action began a chain of events that led to investigation of the accident.

(C9) Many individuals received external and internal radiation exposure exceeding acceptable levels. In total, some 112,000 persons were monitored; 249 persons were contaminated either internally or externally. Some suffered very high internal and external contamination owing to the way they had handled the caesium chloride powder, such as daubing their skin, eating with contaminated hands, and handling various objects. Four persons died within four weeks of admission to hospital, having received total body doses of at least 5 Gy.

(C10) Seven main sites of contamination were identified, including the scrap-yards concerned, some of them with dose rates of up to 2 $Sv.h^{-1}$ at one metre. Aerial surveys were conducted over 67 square kilometres. Of 159 houses monitored, 42 required decontamination. The final volume of waste stored was 3500 cubic metres or 275 lorry loads. The radioactivity accounted for in the decontamination was estimated at 1,200 Ci (44.4 TBq), compared with the known activity of the source before the accident of 1375 Ci (51 TBq).

(C11) This case is similar to that discussed in Section C.1. The following factors contributed to the accident: lack of compliance with regulations in that there was no safe, formal, and complete decommissioning of the facility, and the ^{137}Cs teletherapy unit, no longer in use, was abandoned unsecured.

REFERENCES

AAPM (1983) A protocol for the determination of absorbed dose from high-energy photon and electron beams. Report of the American Association of Physicists in Medicine Radiation Therapy Committee Task Group No. 21. Med. Phys. 10, 741–771.

AAPM (1993) Medical accelerator safety considerations. Report of the American Association of Physicists in Medicine Radiation Therapy Committee Task Group No. 35. Med. Phys. 20, 1261–1275.

AAPM (1994) Comprehensive QA for radiation oncology. Report of the American Association of Physicists in Medicine Radiation Therapy Committee Task Group 40. Med. Phys. 21, 1–38.

AAPM (1998) Quality assurance for clinical radiotherapy treatment planning. Report of the American Association of Physicists in Medicine Radiation Therapy Committee Task Group 53. Med. Phys. 25, 1773–1829.

AAPM (1999) Protocol for clinical reference dosimetry of high-energy photon and electron beams. Report of the American Association of Physicists in Medicine Radiation Therapy Committee Task Group No. 51. Med. Phys. 26, 1847–1870.

Alletti, P., Bey, P. (1996) Recommendations for a Quality Assurance programme in external radiotherapy. ESTRO Booklet 2, Garant, Leuven, Belgium.

Andreo, P. (1990) Uncertainties in dosimetric data and beam calibration. Int. J. Radiation Oncology Biol. Phys. 19, 1233–1247.

Andreo, P. (1992) Absorbed dose beam quality factors for the dosimetry of high-energy photon beams. Phys. Med. Biol. 37, 2189–2211.

Andreo, P., Brosed, A., Gultresa, J., et al. (1997) Tolerance, effective tolerance and action levels in a quality assurance program for equipment used in radiotherapy (in Spanish). In: *Proc. XI Annual Meeting of the Spanish Society of Medical Physics, Valencia, Spain.* SEFM, Madrid, pp. 39–65.

Ash, D., Bates, T. (1994) Report on the clinical effects of inadvertent radiation underdosage in 1045 patients. Clinical Oncology 6, 214–225.

Brahme, A., Chavaudra, J., Landberg, T., et al. (1988). Accuracy Requirements and Quality Assurance of External Beam Therapy with Photons and Electrons Acta Oncol. Suppl. 1.

CNSNS (1985) Accidente por contaminación con ^{60}Co México 1984. CNSNS-IT-001. Comisión Nacional de Seguridad Nuclear y Salvaguardias, México.

Cohen, L., Schultheiss, T.E., Kennaugh, R.C. (1995). A radiation overdose incident: initial data. Int. J. Radiat. Oncol. Biol. Phys. 33, 217–224.

Cosset, J.M, Dubray, B., Girinsky, T., et al. (1994) Fractionation sensitivity of mammalian tissues. Adv. in Radiation Biol. 18, 91–121.

Dische, S., Saunders, M. I., Williams, C., et al. (1993) Precision in reporting the dose in a course of radiotherapy. Radiother. Oncol. 29, 287–293.

Dutreix, A. (1984) When and how can we improve precision in radiotherapy? Radioth. Oncol. 2, 275–292.

Dutreix,A., van der Schueren, E., Derreumaux, S., et al. (1993) Preliminary results of a quality assurance network for radiotherapy centres in Europe. Radiother. Oncol. 29, 97–101.

El Periódico (1993) Account of the court proceedings and verdict for the case of the accelerator accident in Zaragoza (Spain), reproduced in the newspaper of April 7 (in Spanish). El Periódico, Madrid.

ESTRO (1995) European Society for Therapeutic Radiology and Oncology: Quality assurance in radiotherapy. Radiother. Oncol. 35, 61–73.

ESTRO (1996) Quality assurance in radiotherapy: the importance of medical physics staffing levels. Recommendations from an ESTRO/EFOMP joint task group. Radiother. Oncol. 41, 89–94.

Ferreira, I. H., Dutreix, A., Bridier, A., et al. (2000) The ESTRO-QUALity assurance network (EQUAL). Radiother. Oncol. 55, 273–282.

Garavaglia, G., Johansson, K., Leunens, G., et al. (1993) The role of in-vivo dosimetry. Radiother.Oncol. 29, 281–282.

Goitein, M. (1983) Nonstandard deviations. Med. Phys. 10, 709–711.

Hanks, G. (1984) Quality Assurance in Radiation Therapy: Clinical and Physical Aspects. Pergamon Press, New York, NY.

Hanson W.F., Aguirre J.F., Stovall, M. (1997) Radiotherapy physics quality audit networks in the USA. In: *Quality Assurance in Radiotherapy: Proceedings of an IAEA-ISRO working meeting held in Vienna May 8-9, 1995.* IAEA Tec-Doc 989, IAEA, Vienna, Austria, pp. 195–203.

Herring, D.F., Compton, D.M.J. (1971) The degree of precision required in the radiation dose delivered in cancer therapy. In: *Brit. J. Radiol. Special Report No. 4*. British Institute of Radiology, London, UK, pp. 56–78.

Horiot, J.C. (1995). Radiotherapy: still young and almost a hundred years. The Magazine of the World Health Organization. 48th year, No. 3. May/June, 1995.

IAEA (1988) *The radiological accident in Goiania*. STI/PUB/815. International Atomic Energy Agency, Vienna, Austria.

IAEA (1987) Absorbed Dose Determination in Photon and Electron Beams: An International Code of Practice, Technical Report Series, TRS-277. International Atomic Energy Agency, Vienna, Austria.

IAEA (1996) International Basic Safety Standards for Protection against Ionizing Radiation and for the Safety of Radiation Sources, IAEA Safety Series, No. 115. International Atomic Energy Agency, Vienna, Austria.

IAEA (1997) *Quality Assurance in Radiotherapy*, Proceedings of the Working Meeting on National Programmes: Design, Harmonization and Structures, jointly organized by the IAEA and the International Society for Radiation Oncology (ISRO), Vienna 1995. IAEA Tec-Doc 989. International Atomic Energy Agency, Vienna, Austria.

IAEA (1998a) Design and Implementation of a Radiotherapy Programme: Clinical, Medical Physics, Radiation Protection and Safety Aspects. IAEA Tec-Doc 1040. International Atomic Energy Agency, Vienna, Austria.

IAEA (1998b) *Accidental overexposure of radiotherapy patients in San José, Costa Rica*. IAEA Special publication series. International Atomic Energy Agency, Vienna, Austria.

IAEA (2000a) *Lessons learned from accidental exposures in radiotherapy*. IAEA Safety Report Series 117. International Atomic Energy Agency, Vienna, Austria.

IAEA (2000b) Absorbed dose determination in external beam radiotherapy: An international Code of Practice for dosimetry based on standards of absorbed dose to water. IAEA Technical Report Series 398, International Atomic Energy Agency, Vienna, Austria.

ICRP (1991) 1990 Recommendations of the International Commission on Radiological Protection. ICRP Publication 60, Annals of the ICRP 21 (1-3).

ICRP (1997) Protection from potential exposure: Application to selected radiation sources. ICRP Publication 76, Annals of the ICRP 27 (2).

ICRP (1998) Genetic susceptibility to cancer. ICRP Publication 79, Annals of the ICRP 28 (1/2).

ICRU (1976) Determination of absorbed dose in a patient irradiated by beams of X or gamma rays in radiotherapy procedures. ICRU Report 24. International Commission on Radiation Units and Measurements, Bethesda, MD.

ICRU (1985) *Dose and Volume Specification for Reporting Intracavitary Therapy in Gynaecology*, ICRU Report 38. International Commission on Radiation Units and Measurements, Bethesda, MD.

ICRU (1987) *Use of computers in external beam radiotherapy procedures with high-energy photons and electrons*. ICRU Report 42 International Commission on Radiation Units and Measurements, Bethesda, MD.

ICRU (1993) *Prescribing, Recording and Reporting Photon Beam Therapy*, ICRU Report 50. International Commission on Radiation Units and Measurements, Bethesda, MD.

ICRU (1999) *Prescribing, Recording and Reporting Photon Beam Therapy (Supplement to ICRU Report 50)*, ICRU Report 62. International Commission on Radiation Units and Measurements, Bethesda, MD.

IEC (1988) Medical Electrical Equipment- Part 1: General Requirements for Safety. IEC 601-1. International Electrotechnical Commission, Geneva, Switzerland.

IEC (1989) Medical Electrical Equipment- Part 2: Particular Requirements for the Safety of Remote-Controlled Automatically-Driven Gamma-Ray Afterloading Equipment. IEC 60601-2-17. International Electrotechnical Commission, Geneva, Switzerland.

IEC (1993) Medical Electrical Equipment- Part 2: Particular Requirements for the Safety of Radiotherapy Simulators. IEC 60601-2-29. International Electrotechnical Commission, Geneva, Switzerland.

IEC (1997a) Medical Electrical Equipment- Part 2: Particular Requirements for the Safety of Gamma Beam Therapy Equipment. IEC 60601-2-11. International Electrotechnical Commission, Geneva, Switzerland.

IEC (1997b) Medical Electrical Equipment. General Requirements for Safety. 4. Collateral Standard: Programmable Electrical Medical Systems. IEC 60601-1-4. International Electrotechnical Commission, Geneva, Switzerland.

IEC (1998) Medical Electrical Equipment- Part 2: Particular Requirements for the Safety of Electron Accelerators in the Range 1 to 50 MeV. IEC 60601-2-1. International Electrotechnical Commission, Geneva, Switzerland.

IEC (in press) Medical Electrical Equipment- Part 2: Particular Requirements for the Safety of Treatment Planning Systems. IEC 60601-2-xxx. International Electrotechnical Commission, Geneva, Switzerland.

ISCRO (1991) *Radiation Oncology in Integrated Cancer Management (Blue Book)*. Inter-Society Council for Radiation Oncology, Philadelphia, PA.

ISO (1993) International vocabulary of basic and general terms in metrology [Published by ISO in the name of BIPM, IEC, IFCC, IUPAC, IUPAP and OIML]. International Organization for Standardization, Geneva, Switzerland.

ISO (1995) Guide to the expression of uncertainty in measurement, 2nd ed. [Published by ISO in the name of BIPM, IEC, IFCC, IUPAC, IUPAP and OIML]. International Organization for Standardization, Geneva, Switzerland.

ISO (2000) *ISO Standards Compendium ISO 9000- Quality Management, 8th edition*. International Organization for Standardization, Geneva, Switzerland.

Izewska I., Andreo, P. (2000) The IAEA/WHO TLD postal programme for radiotherapy hospitals. Radiother. Oncol. 54, 65–72.

Järvinen, H. (1997) Implementation of the quality assurance programme at the national level. In: Quality Assurance in Radiotherapy: Proceedings of an IAEA-ISRO working meeting held in Vienna May 8-9, 1995, IAEA Tec-Doc 989, IAEA, Vienna, Austria, pp. 205–213.

Kesteloot, K., Dutreix, A., Van der Schueren, E. (1993) A model for calculating the costs of in vivo dosimetry and portal imaging in radiotherapy departments. Radiother. Oncol. 28, 108–117.

Leunens, G., Van Dam, J., Dutreix, A., et al. (1990) Quality Assurance in radiotherapy by in-vivo dosimetry 1. Entrance dose measurements, a reliable procedure. Radiother.Oncol. 17, 141–151.

Loevinger, R., Loftus T.P. (1977) Uncertainty in the delivery of absorbed dose. In: Casnati, E. (Ed.): Ionizing Radiation Metrology, Editrice Compositori, Bologna, Italy, pp 459–473.

Leveson, N., Turner, C.S. (1993) An investigation of the Therac-25 accidents. IEEE Computer, 26 (7), 18–41.

Maciejewski, B., Withers, H.R., Taylor, J.M.G., et al. (1990) Dose fractionation and regeneration in radiotherapy for cancer of the oral cavity and oropharynx. Part 2. Normal tissue responses: acute and late effects. Int. J. Rad. Oncol. Biol. Phys. 18, 101–111.

Mijnheer, B.J. (1996) The clinical basis for dosimetric accuracy in radiotherapy. In: Radiation incidents: Uncertainties, errors and accidents in radiotherapy, a meeting organised by the British Institute of Radiology. British Institute of Radiology, London, UK, pp. 16–20.

Mijnheer, B.J., Battermann, J.J., Wambersie, A. (1987) What degree of accuracy is required and can be achieved in photon and neutron therapy? Radiother. Oncol. 8, 237–252.

Mijnheer, B.J., Battermann, J.J., Wambersie, A. (1989) Reply to Precision and accuracy in radiotherapy (Letter to the Editor). Radiother. Oncol. 14, 163–167.

NRC (1988). Report to Congress on Abnormal Occurences. AS-88-8. Multiple Medical Teletherapy Misadministrations at Sacred Heart Hospital in Cumberland, Maryland. NUREG-0090, Volume 11, No. 4. US Nuclear Regulatory Commission, Washington, DC.

NRC Nuclear Regulatory Commission. (1992). Report to Congress on Abnormal Occurences. 92-18. Loss of Iridium-192 Source and Medical Therapy Misadministration at Indiana Regional Cancer Center in Indiana, Pennsylvania. NUREG-0090, Volume 15, No. 4. Washington DC.

NRC (1996). Report to Congress on Abnormal Occurences. AS-88-8. Multiple Medical Teletherapy Misadministrations at Sacred Heart Hospital in Cumberland, Maryland. NUREG-0090, Volume 19. US Nuclear Regulatory Commission, Washington, DC.

Ortiz, P. (1998) Lessons learned from accidents in radiotherapy. In: International Conference on the Safety of Radiation Sources and the Security of Radioactive Materials, Dijon 1998, IAEA TEDCOC-1045. IAEA, Vienna, Austria.

Pérez, L.A., Brady, L.W. (1992). Principles and Practice of Radiation Oncology. J.B. Lippincott Company. Philadelphia, PA.

Powers, B.E., Thames, H.D., Gillette, S.M., et al. (1998) Volume effects in the irradiated canine spinal cord: do they exist when the probability of injury is low?. Radioth. and Oncol. 46, 207–306.

Rawlinson, J.A. (1987) *Report on the Therac-25.* OCTRF/OCI Physicists Meeting. Kingston, Ont. Canada, May 7, 1987.

Rubin, P., Casarett, G.W. (1968) *Clinical Radiation Pathology.* W.B. Saunders, Philadelphia, PA.

Sebag-Montefiore, D.J., Maher, E.J., Young, J., et. al. (1992) Variation in mantle technigue:implications for establishing priorities for quality assurance in clinical trials. Radiotherapy and Oncology 23 (3), 144–149.

Sharp, C., Cox, R. (1999). Genetic susceptibility to radiation effects: possible implication for medical ionising radiation exposures. Eur. J. Nucl. Med. 26, 425–428.

Svensson, H. (1984) Quality assurance in radiation therapy: physical aspects. Int. J. Radiat. Oncol. Biol. Phys. 10, 59–65.

Thames, H.D., Hendry, J.H. (1987). *Fractionation in Radiotherapy.* Taylor and Francis, London., UK.

Thwaites, D.I. (1994) Uncertainties at the end point of the basic dosimetry chain. In: *Proc. Symp. Measurement Assurance in Dosimetry,* Vienna, 1993. IAEA Proceedings Series STI/PUB/930. IAEA, Vienna, Austria, pp. 239–255.

Thwaites, D., Scalliet, P., Leer, J.W., et al. (1995) Quality Assurance in Radiotherapy. Radiother.Oncol. 35, 61–73.

Van Dam, J., Marinello, G. (1994) *Methods for in-vivo dosimetry in external radiotherapy.* ESTRO Booklet 1, Garant, Leuven, Belgium.

Wambersie, A., Van Dam, J., Hanks, G., et al. (1994). What accuracy is needed in dosimetry? In: *Proc. Sem. Radiation dose in radiotherapy from prescription to delivery,* Leuven, 1991. IAEA-TECDOC-734. IAEA, Vienna, Austria, pp. 11–35.

Wambersie, A., Zreik, H., Prignot, M., et al. (1984) Variation of RBE as a function of depth in a high energy electron beam in the first millimetres of irradiated tissues determined by observation of skin reactions on patients (a clinical trial). Strahlentherapie 148, 279–287.

WHO (1988) *Quality Assurance in Radiotherapy.* World Health Organization, Geneva, Switzerland.

International Commission on Radiological Protection:

History, policies, procedures

A free booklet describing ICRP

The International Commission on Radiological Protection (ICRP) recently issued this document as a straightforward review of its history, mode of operations, concepts, and current policies. After an Introduction, it presents the History and affiliation and The structure and mode of operation of the Commission. The objectives of the Commission's recommendations are laid out; Quantities, The biological basis of the Commission's policy, Concepts, and The quantitative basis for risk estimates are briefly explained. The structure of the system of protection, Problems of interpretation and application in the system of protection, and The need for stability, consistency, and clarity are all briefly discussed.

This booklet is not part of the *Annals of the ICRP*, and is not automatically distributed to subscribers.

Individual copies, in English or French, may be requested free of charge from

The Scientific Secretary, Dr J Valentin
ICRP
SE-171 16 Stockholm
Sweden.

Bulk deliveries (5 or more copies) can be arranged through the ICRP Scientific Secretary, Dr Valentin, at £ 1.00 per copy.

The International Commission on Radiological Protection

Our Mission Statement

The International Commission on Radiological Protection, ICRP, is an independent Registered Charity, established to advance for the public benefit the science of radiological protection, in particular by providing recommendations and guidance on all aspects of protection against ionising radiation.

The primary body in radiological protection is ICRP. It was formed in 1928 as the 'International X-ray and Radium Committee', but adopted its present name in 1950 to reflect its growing involvement in areas outside that of occupational exposure in medicine, where it originated.

ICRP consists of the Main Commission, Committee 1 (Radiation Effects), Committee 2 (Doses from Radiation Exposure), Committee 3 (Protection in Medicine), Committee 4 (Application of ICRP Recommendations), *ad hoc* Task Groups and Working Parties, and the Scientific Secretariat.

The Main Commission consists of 12 members and a Chairman, while the Committees contain between 15 and 20 members each. The Commission and its Committees run for four-year periods, from 1 July. On each occasion of a new period, at least three, and not more than five, members of the Commission must be changed. A similar rate of renewal is sought for the Committees. Such a new period began 1 July 1997.

The Commission meets once or twice a year. Each Committee meets once a year. Twice in each four-year period, the annual meeting of the Committees is conducted jointly and together with the Commission. These meetings are funded as necessary from monies available to ICRP.

The activities of ICRP are financed mainly by voluntary contributions from national and international bodies with an interest in radiological protection. Some additional funds accrue from royalties on ICRP *Publications*. Members' institutions also provide support to ICRP by making the members' time available without charge and, in many cases, contributing to their costs of attending meetings.

The Commission uses Task Groups and Working Parties to deal with specific areas. Task Groups are formally appointed by the Commission to perform a defined task, usually the preparation of a draft report. A Task Group usually contains a majority of specialists from outside the Commission's structure. It is funded as necessary from monies available to ICRP.

Working Parties are set up by Committees to develop ideas, sometimes leading to the establishment of a Task Group. The membership of a Working Party is usually limited to Committee members. Working Parties receive no funding of their own, *i.e.* they operate primarily by correspondence and by meetings in direct conjunction with meetings of the Committee concerned.

These activities are co-ordinated with a minimum of bureaucracy by a Scientific Secretary, ensuring that ICRP recommendations are promulgated.

Thus, ICRP is an independent international network of specialists in various fields of radiological protection. At any one time, about 100 eminent scientists are actively involved in the work of ICRP. The four-tier structure described provides a rigorous Quality Management system of peer review for the production of ICRP Publications.

In preparing its recommendations, the Commission considers the fundamental principles and quantitative bases on which appropriate radiation protection measures can be established, while leaving to the various national protection bodies the responsibility of formulating the specific advice, codes of practice, or regulations that are best suited to the needs of their individual countries. The aim of the recommendations of ICRP is to

—provide an appropriate standard of protection for mankind from sources of ionising radiation, without unduly limiting beneficial practices that give rise to exposure to radiation.

Composition of the International Commission on Radiological Protection and Committees, 1997–2001

MAIN COMMISSION
Chairman:
Professor R H Clarke

Vice-Chairman:
Dr L-E Holm

Members:
Professor R Alexakhin
Dr J D Boice Jr.
Dr R Cox
Ms G J Dicus
Dr A J González
Professor F A Mettler
Dr Y Sasaki
Professor C Streffer
Dr A Sugier
Professor Z Q Pan
Mr B C Winkler

Emeritus Members: (as of July 2001)
Mr H J Dunster
Professor B Lindell
Dr W K Sinclair
Dr L S Taylor

Scientific Secretary
Dr J Valentin

COMMITTEE 1
Chairman:
Dr R Cox

Members:
Dr A V Akleyev
Professor M Blettner
Dr J H Hendry
Professor Dr A M Kellerer
Dr C Land
Professor J Little
Dr C Muirhead
Dr O Niwa
Dr D Preston
Dr R J Preston
Dr E Ron
Professor K Sankaranarayanan
Professor R E Shore
Dr F Stewart
Dr M Tirmarche
Dr R Ullrich
Dr P-K Zhou

COMMITTEE 2

Chairman:
Professor C Streffer

Members
Dr M Balonov
Dr B Boecker
Dr A Bouville
Professor Dr G Dietze
Dr K F Eckerman
Miss F A Fry
Dr J Inaba
Professor Dr I A Likhtarov
Professor J L Lipsztein
Dr H G Menzel
Dr H Métivier
Dr H Paretzke
Dr A S Pradhan
Dr J W Stather
Professor D M Taylor
Dr Y Zhou

COMMITTEE 3
Chairman:
Professor F Mettler

Members:
Professor J M Cosset
Dr C Cousins
Dr M J Guiberteau
Dr I Gusev
Dr L K Harding
Dr M Hiraoka
Professor Dr J Liniecki
Professor S Mattsson
Dr P Ortiz-Lopez
Dr L V Pinillos-Ashton
Professor M M Rehani
Professor H Ringertz
Dr M Rosenstein
Dr C Sharp
Professor W Yin

COMMITTEE 4
Chairman:
Mr B C Winkler

Members:
Dr E d'Amato
Dr D Cancio
Dr M E Clark
Dr D Cool
Dr J Cooper
Dr T Kosako
Dr J F Lecomte
Mr J Lochard
Dr G C Mason
Dr A C McEwan
Dr M Measures
Mr M Savkin
Dr J E Till
Dr K Ulbak
Dr W Weiss
Dr Y Xia
Dr C Zuur

In compliance with GPSR, should you have any concern about the safety of this product, please advise: International Associates Auditing and Certification Limited, The Black Church, St Mary's Place, Dublin 7, D07 P4AX Ireland EUAR@ie.ia-net.com

Printed and bound by CPI Group (UK) Ltd, Croydon, CR0 4YY

12/06/2026

02136155-0001